Die-stacking Architecture

Synthesis Lectures on Computer Architecture

Editor
Margaret Martonosi, *Princeton University*

Synthesis Lectures on Computer Architecture publishes 50- to 100-page publications on topics pertaining to the science and art of designing, analyzing, selecting and interconnecting hardware components to create computers that meet functional, performance and cost goals. The scope will largely follow the purview of premier computer architecture conferences, such as ISCA, HPCA, MICRO, and ASPLOS.

Die-stacking Architecture
Yuan Xie and Jishen Zhao
2015

Power-Efficient Computer Architectures: Recent Advances
Magnus Själander, Margaret Martonosi, and Stefanos Kaxiras
2014

FPGA-Accelerated Simulation of Computer Systems
Hari Angepat, Derek Chiou, Eric S. Chung, and James C. Hoe
2014

A Primer on Hardware Prefetching
Babak Falsafi and Thomas F. Wenisch
2014

On-Chip Photonic Interconnects: A Computer Architect's Perspective
Christopher J. Nitta, Matthew K. Farrens, and Venkatesh Akella
2013

Optimization and Mathematical Modeling in Computer Architecture
Tony Nowatzki, Michael Ferris, Karthikeyan Sankaralingam, Cristian Estan, Nilay Vaish, and David Wood
2013

Security Basics for Computer Architects
Ruby B. Lee
2013

Die-stacking Architecture

Yuan Xie and Jishen Zhao

ISBN: 978-3-031-00619-7   paperback
ISBN: 978-3-031-01747-6   ebook

DOI 10.1007/978-3-031-01747-6

A Publication in the Springer series
SYNTHESIS LECTURES ON COMPUTER ARCHITECTURE

Lecture #31
Series Editor: Margaret Martonosi, *Princeton University*
Series ISSN
Print 1935-3235   Electronic 1935-3243

Die-stacking Architecture

Yuan Xie
University of California, Santa Barbara

Jishen Zhao
University of California, Santa Cruz

SYNTHESIS LECTURES ON COMPUTER ARCHITECTURE #31

ABSTRACT

The emerging three-dimensional (3D) chip architectures, with their intrinsic capability of reducing the wire length, promise attractive solutions to reduce the delay of interconnects in future microprocessors. 3D memory stacking enables much higher memory bandwidth for future chip-multiprocessor design, mitigating the "memory wall" problem. In addition, heterogenous integration enabled by 3D technology can also result in innovative designs for future microprocessors. This book first provides a brief introduction to this emerging technology, and then presents a variety of approaches to designing future 3D microprocessor systems, by leveraging the benefits of low latency, high bandwidth, and heterogeneous integration capability which are offered by 3D technology.

KEYWORDS

emerging technology, die-stacking, 3D integrated circuits, memory architecture, heterogeneous integration

Contents

Preface

Three-dimensional (3D) integration is an emerging technology, where two or more layers of active devices (e.g., CMOS transistors) are integrated both vertically and horizontally in a single circuit. With continuous technology scaling, 3D integration is becoming an increasingly attractive technology in implementing microprocessor systems by offering much lower power consumption, lower interconnect latency, and higher interconnect bandwidth compared to traditional two-dimensional (2D) circuit integration.

In particular, 3D integration technologies promise at least four major benefits toward future microprocessor design.

- **Reduced interconnect wire length.** 3D integration can dramatically reduce interconnect wire length, especially by reducing global interconnects. This can directly lead to two benefits: improved circuit delay and reduced power consumption. Circuit delay reduction can be a straightforward effect of the reduced wire length; it can result in substantial system performance improvement. The power reduction is a result of reduced parasitic capacitance due to the shorter wire length and incorporating previously off-chip signals to be on-chip; it can lead to less heat generation and extended battery life.

- **Improved memory bandwidth.** 3D integration can improve memory bandwidth by an order of magnitude, because the number of interconnects between processor and the memory is no longer constrained by off-chip pin counts. In the era of big data and multithreading, applications adopt large memory working sets and multiple threads that simultaneously access the memory. Today, memory bandwidth is a fundamental performance bottleneck. 3D integration can be a critical technology in overcoming the memory bandwidth issue.

- **Enabling heterogeneous integration.** 3D integration enables heterogeneous integration, because circuit layers can be implemented with different and incompatible process technologies. Such heterogeneous integration can lead to novel architecture designs. For example, various incompatible memory technologies, such as SRAM, spin-transfer torque RAM (STT-RAM), and resistive RAM (ReRAM), can be integrated in a single processor chip to form a hybrid cache hierarchy [1].

- **Enabling smaller form factor.** 3D integration enables a much smaller form factor compared to traditional 2D integration technologies. Due to the addition of a third dimension to conventional 2D layout, it leads to a higher packing density and smaller footprint. This potentially leads to processor designs with lower cost.

Both academia and the semiconductor industry are actively pursuing this technology by developing efficient architectures in a variety of forms. From the industry prospective, 3D integrated memory is envisioned to become pervasive in the near future. Intel's Xeon Phi processors will deliver with 3D integrated DRAMs in 2016 [2]. NVIDIA announced that 3D integrated memory will be adopted in their new GPU products in 2016 [3]. AMD plans to ship high-bandwidth memory (HBM) with their GPU products and heterogeneous system architecture (HSA)-based CPUs in 2015 [4]. From the academia prospective, comprehensive studies have been performed across all aspects of microprocessor architecture design by employing 3D integration technologies, such as 3D stacked processor core and cache architectures, 3D integrated memory, and 3D network-on-chip. Furthermore, a large body of research has studied critical issues and opportunities raised by adopting 3D integration technologies, such as thermal issues which are imposed by dense integration of active electronic devices, cost issues which are incurred by extra process and increased die area, and the opportunity in designing cost-effective microprocessor architectures.

This book provides a detailed introduction to architecture design with 3D integration technologies. The book will start with presenting the background of 3D integration technologies (Chapter 1), followed by a detailed analysis of the benefits offered by these technologies including low latency, high bandwidth, heterogeneous integration capability, and cost efficiency (Chapter 2). Then, it will review various approaches to designing future 3D integrated microprocessors by leveraging the benefits of 3D integration (Chapter 3 through Chapter 6). These approaches cover all levels of microprocessor systems, including processor cores, caches, main memory, and on-chip network. Furthermore, this book discusses thermal issues raised by 3D integration and presents recently proposed thermal-aware architecture designs (Chapter 7). Finally, this book presents a comprehensive cost model which is built based on detailed cost analysis for fabricating 3D integrated microprocessors (Chapter 8). By utilizing the cost model, the book presents and compares cost-effective microprocessor design strategies.

While this book mostly focuses on designing high-performance processors, the concepts and techniques can also be applied to other market segments such as embedded processors and exascale high-performance computing (HPC) systems.

The target audiences for this book are students, researchers, and engineers in IC design and computer architecture, who are interested in leveraging the benefits of 3D integration for their designs and research.

Yuan Xie and Jishen Zhao
June 2015

Acknowledgments

Much of the work and ideas presented in this book have evolved over years in working with our colleagues and graduate students at Pennsylvania State University (in particular Professor Vijaykrishnan Narayanan, Professor Mary Jane Irwin, Professor Chita Das), and our industry collaborators including Dr. Gabriel Loh, Dr. Bryan Black, Dr. Norm Jouppi, and Mr. Kerry Bernstein.

We also thank Prof. Niraj Jha, Prof. Margaret Martonosi, and other reviewers for the comments and feedback to improve the draft.

Yuan Xie and Jishen Zhao
June 2015

CHAPTER 1

3D Integration Technology

A 3D integrated circuit (3D IC) has two or more active device layers (i.e., CMOS transistor layers) integrated vertically as a single chip, using various integration methods. This chapter will give a brief introduction to different 3D integration technologies, including monolithic 3D ICs and through-silicon-via (TSV)-based 3D ICs.

1.1   3D INTEGRATED CIRCUITS VS. 3D PACKAGING

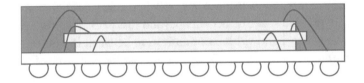

Figure 1.1: Illustration of 3D SiP (system-in-package) technology.

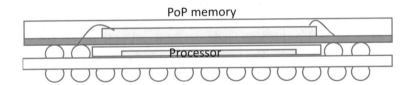

Figure 1.2: Illustration of 3D PoP (package-on-package) technology.

Even though the 3D integrated circuit is an emerging technology, the 3D packaging is a mature technology and has been widely adopted by industry. There are two main stream 3D packaging methods:

- System-in-package (SiP) stacks separate IC chips together vertically on a substrate in a single package, with internal wire-bond connections that are bonded to the package. The conceptual view of SiP is shown in Figure 1.1.

- Package-on-package (PoP) stacks separate packages vertically. The conceptual view of PoP is shown in Figure 1.2.

There are many commercial products that use 3D packaging. For example, Apple A8 processor (used in iPhone6) integrates a processor package (that has dual-core CPU and quad-core GPU) with a 1GB of LPDDR3 DRAM package using package-on-package technology. The 3D chip stacking based on SiP or PoP is packaging technology with the design goal of saving space. The chips in the stack communicate with off-chip signaling, which usually has limited connectivity and long delay. It does not require significant change in processor architecture or design methodologies. On the contrary, the emerging 3D integrated circuits can provide a large number of connections (and much faster) between two stacked dies, and therefore enables revolutionary improvement in architecture innovation and requires fundamental changes in design methodologies.

1.2   DIFFERENT PROCESS TECHNOLOGIES FOR 3D ICS

Different from SiP or PoP, which uses off-chip signaling for communication between stacked dies, the emerging 3D integrated technologies develop fine-granularity interconnects between layers, and therefore provide superior benefits beyond just packaging for space saving, including fast interconnects and high-bandwidth communication between layers.

The 3D integration technologies [5, 6] can be classified into one of the two following categories:

- *Monolithic approach.* This approach involves a sequential device process. The front-end processing (to build the device layer) is repeated on a single wafer to build multiple active device layers before the back-end processing builds interconnects among devices.

- *Stacking approach,* which could be further categorized as wafer-to-wafer, die-to-wafer, or die-to-die stacking methods. This approach processes each layer separately, using conventional fabrication techniques. These multiple layers are then assembled to build up 3D IC, using bonding technology. Since the stacking approach does not require the change of conventional fabrication process, it is much more practical compared to the monolithic approach, and become the focus of recent 3D integration research.

Several 3D stacking technologies have been explored recently, including wire bonded, microbump, contactless (capacitive or inductive), and *through-silicon vias (TSV)* vertical interconnects. Among all these integration approaches, TSV-based 3D integration has the potential to offer the greatest vertical interconnect density, and therefore is the most promising one among all the vertical interconnect technologies. Figure 1.3 shows a conceptual 2-layer 3D integrated circuit with TSV and microbump.

3D stacking can be implemented using two major techniques [7]: (1) *Face-to-Face (F2F)* bonding: two wafers (dies) are stacked so that the very top metal layers are connected. Note that the die-to-die interconnects in face-to-face wafer bonding does not go through a thick buried silicon layer and can be fabricated as *microbump*. The connections to C4I/O pads[1] (which are chip pads used to mount the chip to external circuitry) are formed as TSVs; (2) *Face-to-Back (F2B)* bonding: multiple device layers are stacked together with the top metal layer of one die bonded together with the substrate of the other die, and direct vertical interconnects (which are called *through-silicon vias (TSV)*) tunneling through the substrate. In such F2B bonding, TSVs are used for both between-layer-connections and I/O connections. Figure 1.3 shows a conceptual 2-layer 3D IC with F2F or F2B bonding, with both TSV connections and microbump connections between layers.

All TSV-based 3D stacking approaches share the following three common process steps [7]:

- *TSV formation.* This is the step to fabricate the through-silicon-via on a wafer. It can also be done in 3 different ways: (1)*Via-first*: The TSVs are built before any CMOS transistors are fabricated on the wafer; (2)*Via-middle*: The TSVs. are built after the CMOS transistors are fabricated but before the metal layers are fabricated; (3)*Via-last*: The TSVs are built after both CMOS transistors and metal connections among transistors are fabricated. Via-middle can only be done in a foundry, while the other two approaches can be done outside a foundry. Currently there are two types of materials: highly conductive copper TSVs, or smaller Tungsten TSVs.

- *Wafer thinning.* Wafer thinning is used to reduce the overheads of TSVs. Since TSVs. need to maintain a certain aspect ratio for reliability/manufacturability purpose, it is a critical step to thin the wafer so that we can build small and short TSVs between layers. The thinner the wafer, the smaller (and shorter) the TSV is (with the same aspect ratio constraint) [7]. The wafer thickness could be in the range of 10 μm to 100 μm and the TSV size is in the range of 0.2 μm to 10 μm [8].

- *Aligned wafer and die bonding*, which could be wafer-to-wafer (W2W) bonding or die-to-wafer (D2W) bonding.

1.3   THE IMPACT OF 3D TECHNOLOGY ON 3D MICROPROCESSOR PARTITIONING

One of the key issues for 3D IC design is what level of logic granularity should be considered in 3D partitioning: designers may perform a fine-granularity partitioning at the logic gate level, or perform a coarse-granularity partitioning at the core level (such as keeping the processor core to be a 2D design, with cache stacked on top of the core). Which partitioning strategy should be adopted is heavily influenced by the underlying 3D process technology.

---

[1]C4 is the acronym of Controlled Collapse Chip Connection, also known as Flip Chip, a method for interconnecting IC chips to external circuitry with the chip IO pads.

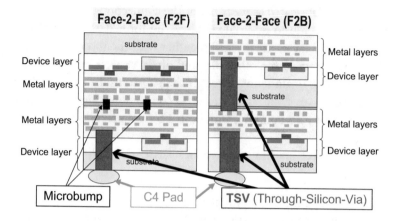

Figure 1.3: Illustration of F2F and F2B 3D bonding.

Level	Area (mm2)	Number of interconnects	Interconnects per mm2
Gate	4×10^{-6}	1	2.5×10^5
Macro	0.04	200	5000
Unit	4	2000	500
Core	40	2000	50

Figure 1.4: Comparison of interconnections at different partitioning granularity [9].

In TSV-based 3D stacking bonding, the dimension of the TSVs. is not expected to scale at the same rate as feature size because alignment tolerance during bonding poses limitations on the scaling of the vias. The TSV size, length, and the pitch density, as well as the bonding method (face-to-face or face-to-back bonding), can have a significant impact on the 3D IC design. For example, a relatively large size of TSVs. can hinder partitioning a design at very fine granularity across multiple device layers, and make the true 3D component design less possible. On the other hand, monolithic 3D integration provides more flexibility in vertical 3D connection because the vertical 3D via can potentially scale down with feature size due to the use of local wires for connections. Availability of such technologies makes it possible to partition the design at a very fine granularity. Furthermore, face-to-face bonding or SOI-based 3D integration may have a smaller via pitch size and higher via density than face-to-back bonding or bulk-CMOS-based integration. Such influence of the 3D technology parameters on the microprocessor design must be thoroughly studied before an appropriate partition strategy is adopted.

With TSV-based 3D stacking, the partitioning strategy is determined by the TSV pitch size and the via diameter. As shown in Figure 1.3, the TSV goes through the substrate and incurs

		Via Diameter				
		0.5	1	5	10	20
Partition Level	gate	5	20	500	2000	8000
	macro	0.12	0.5	12	50	200
	unit	0.01	0.05	1.2	2	20
	core	0.001	0.01	0.12	0.5	2

Figure 1.5: Area overhead (in relative ratio, via diameter in unit of μm) at different partitioning granularity [9].

area overhead, and therefore the larger the via diameter, the higher the area overhead. For example, Figure 1.4 shows the number of connections for different partitioning granularity, and Figure 1.5 shows the area overhead for different size of 3D via diameters.[2] They show that for fine granularity partitioning, there are a lot of connections, and with relative large via diameter, the area overhead would be very high for fine granularity partitioning. Consequently, for most of the existing 3D process technology with via diameter usually larger than 1 μm, it makes more sense to perform the 3D partitioning at the unit level or core level, rather than at the gate level that can result in a large area overhead.

---

[2]These two tables are based on IBM 65nm technology for high performance microprocessor design.

CHAPTER 2

Benefits of 3D Integration

The following subsections will discuss various architecture design approaches that leverage different benefits that 3D integration technology can offer, namely, wire length reduction, high memory bandwidth, heterogeneous integration, and cost reduction. It will also briefly review 3D network-on-chip architecture designs.

2.1 WIRE LENGTH REDUCTION

Designers have resorted to technology scaling to improve microprocessor performance. Although the size and switching speed of transistors benefit as technology feature sizes continue to shrink, global interconnect wire delay does not scale accordingly with technologies. The increasing wire delays have become one major impediment for performance improvement.

Three-dimensional integrated circuits (3D ICs) are attractive options for overcoming the barriers in interconnect scaling, thereby offering an opportunity to continue performance improvements using CMOS technology. Compared to a traditional two dimensional chip design, one of the important benefits of a 3D chip over a traditional two-dimensional (2D) design is the reduction on global interconnects. It has been shown that three-dimensional architectures reduce wiring length by a factor of the square root of the number of layers used [10]. The reduction of wire length due to 3D integration can result in two obvious benefits: *latency improvement* and *power reduction*.

Latency Improvement.   Latency improvement can be achieved due to the reduction of average interconnect length and the critical path length.

Early work on fine-granularity 3D partitioning of processor components shows that the latency of a 3D component could be reduced. For example, since interconnects dominate the delay of cache accesses which determines the critical path of a microprocessor, and the regular structure and long wires in a cache make it one of the best candidates for 3D designs, 3D cache design is one of the early design examples for fine-granularity 3D partition [5]. Wordline partitioning and bitline partitioning approaches divide a cache bank into multiple layers and reduce the global interconnects, resulting in a fast cache access time. Depending on the design constraints, the 3DCacti tool [11] automatically explores the design space for a cache design, and identifies the optimal partitioning strategy; the latency reduction can be up to 25% for a two-layer 3D cache. 3D arithmetic-component designs also show latency benefits. For example, various designs [12–15] have shown that the 3D arithmetic unit design can achieve around 6%–30% delay reduction due to wire length reduction. Such fine-granularity 3D partitioning was also demonstrated by Intel [16], showing that by targeting the heavily pipelined wires, the pipeline modifications resulted in approximately 15% improved performance, when the Intel Pentium-4 processor was folded onto 2-layer 3D implementation.

Note that such fine-granularity design of 3D processor components increases the design complexity, and the latency improvement varies depending on the partitioning strategies and the underlying 3D process technologies. For example, for the same Kogge-Stone adder design, a partitioning based on logic level [12] demonstrates that the delay improvement diminishes as the number of 3D layers increases; a bit-slicing partitioning [14] strategy would have better scalability as the bit-width or the number of layers increases. Furthermore, the delay improvement for such bit-slicing 3D arithmetic units is about 6% when using a bulk-CMOS-based 180nm 3D process [15], while the improvement could be as much as 20% when using a SOI-based 180nm 3D process technology [14], because the SOI-based process has much smaller and shorter TSVs (and therefore much smaller TSV delay) compared to the bulk-CMOS-based process.

Power Reduction.   Interconnect power consumption becomes a large portion of the total power consumption as technology scales. The reduction of the wire length translates into the power saving in 3D IC design. For example, 7% to 46% of power reduction for 3D arithmetic units were demonstrated in [14]. In the 3D Intel Pentium-4 implementation [16], because of the reduction in long global interconnects, the number of repeaters and repeating latches in the implementation is reduced by 50%, and the 3D clock network has 50% less metal RC than the 2D design, resulting in a better skew, jitter, and lower power. Such 3D stacked redesign of Intel Pentium 4 processor improves performance by 15% and reduces power by 15% with a temperature increase of 14 degrees. After using voltage scaling to lower the leak temperature to be the same as the baseline 2D design, their 3D Pentium 4 processor still showed a performance improvement of 8%.

2.2   MEMORY BANDWIDTH IMPROVEMENT

It has been shown that circuit limitations and limited instruction level parallelism will diminish the benefits of modern superscalar microprocessors by increased architectural complexity, which leads to the advent of Chip Multiprocessors (CMP) as a viable alternative to the complex superscalar architecture. The integration of multi-core or many-core microarchitecture on a single die is expected to accentuate the already daunting memory-bandwidth problem. Supplying enough data to a chip with a massive number of on-die cores will become a major challenge for performance scalability. Traditional off-chip memory will not suffice due to the I/O pin limitations. Three-dimensional integration has been envisioned as a solution for future micro-architecture design (especially for multi-core and many-core architectures), to mitigate the interconnect crisis and the "memory wall" problem [17–19]. It is anticipated that memory stacking on top of logic would be one of the early commercial uses of 3D technology for future chip-multiprocessor design, by providing improved memory bandwidth for such multi-core/many-core microprocessors. In addition, such approaches of memory stacking on top of core layers do not have the design complexity problem as demonstrated by the fine-granularity design approaches, which require re-designing all processor components for wire length reduction (as discussed in Sec. 2.1).

Intel [16] explored the memory bandwidth benefits using a base-line Intel Core2 Duo processor, which contains two cores. By having memory stacking, the on-die cache capacity is increased, and the performance is improved by capturing larger working sets, reducing off-chip memory bandwidth requirements. For example, one option is to stack an additional 8MB L2 cache on top of the base-line 2D processor (which contains 4MB L2 cache), and the other option is to replace the SRAM L2 cache with a denser DRAM L2 cache stacking. Their study demonstrated that a 32MB 3D stacked DRAM cache can reduce the cycles per memory access by 13% on average and as much as 55% with negligible temperature increases.

The PicoServer project [20] follows a similar approach to stack DRAM on top of multi-core processors. Instead of using stacked memory as a larger L2 cache (as shown by Intel's work [16]), the fast on-chip 3D stacked DRAM main memory enables wide low-latency buses to the processor cores and eliminates the need for an L2 cache, whose silicon area is allocated to accommodate more cores. Increasing the number of cores by removing the L2 cache can help improve the computation throughput, while each core can run at a much lower frequency, and therefore result in an energy-efficient many core design. For example, the PicoServer design can achieve a 14% performance improvement and 55% power reduction over a baseline multi-core architecture.

As the number of the cores on a single die increases, such memory stacking becomes more important to provide enough memory bandwidth for processor cores. Recently, Intel [21] demonstrated an 80-tile terascale chip with network-on-chip. Each core has a local 256KB SRAM memory (for data and instruction storage) stacked on top of it. TSVs provide a bandwidth of 12GB/second for each core, with a total about 1TB/second bandwidth for Tera Flop computation. In this chip, the thin memory die is put on top of the CPU die, and the power and I/O signals go through memory to CPU.

Since DRAM is stacked on top of the processor cores, the memory organization should also be optimized to fully take advantages of the benefits that TSVs offer [17, 22]. For example, the numbers of ranks and memory controllers are increased, in order to leverage the memory bandwidth benefits. A multiple-entry row buffer cache is implemented to further improve the performance of the 3D main memory. Comprehensive evaluation shows that a 1.75x speedup over commodity DRAM organization is achieved [17]. In addition, MSHR design was modified to provide a scalable L2 miss handling before accessing the 3D stacked main memory. A data structure called the Vector Bloom Filter with dynamic MSHR capacity tuning is proposed. Such structure provides an additional 17.8% performance improvement. If stacked DRAM is used as the last-level caches (LLC) in chip multiple processors (CMPs), the DRAM cache sets are organized into multiple queues [22]. A replacement policy is proposed for the queue-based cache to provide performance isolation between cores and reduce the lifetimes of dead cache lines. Approaches are also proposed to dynamically adapt the queue size and the policy of advancing data between queues.

The latency improvement due to 3D technology can also be demonstrated by such memory stacking design. For example, Li et al. [23] proposed a 3D chip multiprocessor design using network-in-memory topology. In this design, instead of partitioning each processor core or memory bank into multiple layers (as shown in [5, 11]), each core or cache bank remains to be a 2D design. Communication among cores or cache banks are via the network-on-chip (NoC) topology. The core layer and the L2 cache layer are connected with TSV-based bus. Because of the short distance between layers, TSVs provide a fast access from one layer to another layer, and effectively reduce the cache access time because of the faster access to cache banks through TSVs.

2.3   HETEROGENOUS INTEGRATION

3D integration also provides new opportunities for future architecture design, with a new dimension of design space exploration. In particular, the heterogenous integration capability enabled by 3D integration gives designers new perspective when designing future CMPs.

3D integration technologies provide feasible and cost-effective approaches for integrating architectures composed of heterogeneous technologies to realize future microprocessors targeted at the "More than Moore" technology projected by ITRS. 3D integration supports heterogeneous stacking because different types of components can be fabricated separately, and layers can be implemented with different technologies. It is also possible to stack optical device layers or non-volatile memories (such as magnetic RAM (MRAM) or phase-change memory (PCRAM)) on top of microprocessors to enable cost-effective heterogeneous integration. The addition of new stacking layers composed of new device technology will provide greater flexibility in meeting the often conflicting design constraints (such as performance, cost, power, and reliability), and enable innovative designs in future microprocessors.

Non-volatile Memory Stacking.   Stacking layers of non-volatile memory technologies such as Magnetic Random Access Memory (MRAM) [24] and Phase Change Random Access Memory (PRAM) [25] on top of processors can enable a new generation of processor architectures with unique features. There are several characteristics of MRAM and PRAM architectures that make them promising candidates for on-chip memory. In addition to their non-volatility, they have zero standby power, low access power, and are immune to radiation-induced soft errors. However, integrating these nonvolatile memories along with a logic core involves additional fabrication challenges that need to be overcome. For example, the MRAM process requires growing a magnetic stack between metal layers. Consequently, it may incur extra cost and additional fabrication complexity to integrate MRAM with conventional CMOS logic into a single 2D chip. The ability to integrate two different wafers developed with different technologies using 3D stacking offers an ideal solution to overcome this fabrication challenge and exploit the benefits of PRAM and MRAM technologies. For example, Sun et al. [26] demonstrated that the optimized MRAM L2 cache on top of a multi-core processor can improve performance by 4.91% and reduce power by 73.5% compared to the conventional SRAM L2 cache with the similar area.

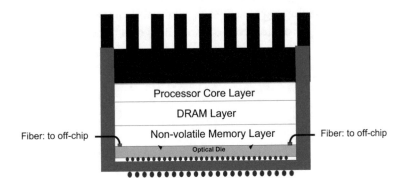

Figure 2.1: An illustration of 3D heterogeneous architecture with non-volatile memory stacking and optical die stacking.

Optical Device Layer Stacking.   Even though 3D memory stacking can help mitigate the memory bandwidth problem, when it comes to off-chip communication, the pin limitations, the energy cost of electrical signaling, and the non-scalability of chip-length global wires are still significant bandwidth impediments. Recent developments in silicon nanophotonic technology have the potential to meet the off-chip communication bandwidth requirements at acceptable power levels. With the heterogeneous integration capability that 3D technology offers, one can integrate optical die together with CMOS processor dies. For example, HP Labs proposed a Corona architecture [27], which is a 3D many-core architecture that uses nanophotonic communication for both inter-core communication and off-stack communication to memory or I/O devices. A photonic crossbar fully interconnects its 256 low-power multithreaded cores at 20 terabyte per second bandwidth, with much lower power consumption.

Figure 2.1 illustrates such a 3D heterogenous processor architecture, which integrates non-volatile memories and optical die together through 3D integration technology.

2.4   COST-EFFECTIVE ARCHITECTURE

Increasing integration density has resulted in large die size for microprocessors. With a constant defect density, a larger die typically has a lower yield. Consequently, partitioning a large 2D microprocessor to be multiple smaller dies and stacking them together may result in a much higher yield for the chip, even though 3D stacking incurs extra manufacture cost due to extra steps for 3D integration and may cause a yield loss during stacking. Depending on the original 2D microprocessor die size, it may be cost-effective to implement the chip using 3D stacking [28], especially for large microprocessors. The heterogenous integration capability that 3D provides can also help reduce the cost.

In addition, as technology feature size scales to reach the physics limits, it has been predicted that moving to the next technology node is not only difficult but also prohibitively expensive. 3D stacking can potentially provide a cost-effective integration solution, compared to traditional technology scaling.

CHAPTER 3

Fine-granularity 3D Processor Design

As 3D integration technology emerges, the 3D stacking provides great opportunities of improvements in the microarchitecture. In this chapter, we introduce some recent 3D research in the architecture level. These techniques leverage the advantages of 3D and help to improve performance, reduce power consumption, etc.

As 3D integration can reduce the wire length, it is straightforward to partition the structure of a planar processor and stack them to improve the performance. There are two different methods: (1) coarse granularity stacking, also known as "memory+logic" strategy, in which some on-chip memories are separated from and stacked on the part containing logic components [29–33], and (2) fine granularity stacking, in which various function units of the processor are separated and stacked, and some components are internally partitioned and implemented with 3D integration [12, 34–37].

3D stacking enables denser form factor and more cost-efficient integration of heterogeneous process technologies. It is possible to stack DRAM or other emerging memories on-chip. Consequently, the advantages of different memory technologies are leveraged. For example, more data can be stored on-chip and the static power consumption can be reduced. At the same time, the high bandwidth provided by 3D integration can be explored for the large capacity on-chip memory to further improve the performance [25, 30–33, 38].

In this chapter, we first describe how to partition and stack the on-chip memory (SRAM arrays), as they have regular structures. Then, we discuss the benefits and issues of partitioning the logic components.

3.1 3D CACHE PARTITIONING

The regular structure and long wires in a cache make it one of the best candidates for 3D designs. This section examines possible partitioning techniques for caches designed using 3D structures and presents a delay and energy model to explore different options for partitioning a cache across different device layers.

3.1.1 3D CACHE PARTITIONING STRATEGIES

In this section, we discuss different approaches to partition a cache into multiple device layers.

The finest granularity of partitioning a cache is at the SRAM cell level. At this level of partitioning, any of the six transistors of a SRAM cell can be assigned to any layer. For example, the pull-up PMOS transistors can be in one device layer and the access transistors and the pull-down NMOS transistors can be in another layer. The benefits of cell level partitioning include the reduction in footprint of the cache arrays and, consequently, the routing distance of the global signals. However, the feasibility of partitioning at this level is constrained by the 3D via size as compared to the SRAM cell size. Assuming a limitation that the aspect ratio of 3D via size cannot be scaled less than $1 \ \mu m \times 1 \ \mu m$, the 3D via has a comparable size to that of a 2D 6T SRAM cell in 180 nm technology and is much larger than a single cell in 70 nm technology. Consequently, when the 3D via size does not scale with feature size (which is the case for wafer-bonding 3D integration), partitioning at the cell level is not feasible. In contrast, partitioning at the SRAM cell level is feasible in technologies such as MLBS, because no limitation is imposed on via scaling with feature size. However, it should be noted that even if the size of a 3D via can be scaled to as small as a nominal contact in a given technology, the total SRAM cell area reduction (as compared to a 2D design) due to the use of additional layers is limited, because metal routing and contacts occupy a significant portion of the 2D SRAM cell area [39]. Consequently, partitioning at a higher level of granularity is more practical. For example, individual sub-arrays in the 2D cache can be partitioned across multiple device layers. The partitioning at this granularity reduces the footprint of cache arrays and routing lengths of global signals. However, it also changes the complexity of the peripheral circuits. In this section, we consider two options of partitioning the sub-array into multiple layers: 3D divided wordline (3DWL) strategy and 3D divided bit line strategy (3DBL).

3D Divided Wordline (3DWL) In this partitioning strategy, the wordlines in a sub-array are divided and mapped onto different active device layers (see Fig. 3.3). The corresponding local wordline decoder of the original wordline in 2D sub-array is placed on one layer and is used to feed the wordline drivers on different layers through the 3D vias. We duplicate wordline drivers for each layer. The duplication overhead is offset by the resized drivers for a smaller capacitive load on the partitioned word line. Further, the delay time of pulling a wordline decreases as the number of pass transistors connected to a wordline driver is smaller. The delay calculation of the 3DWL also accounts for the 3D via area utilization. The area overhead due to 3D vias is small compared to the number of cells on a word line.

Another benefit from 3DWL is that the length of the address line from periphery of the core to the wordline decoder decreases in proportion to the number of device layers. Similarly, the routing distance between the output of pre-decoder to the local decoder is reduced. The select lines for the writes and multiplexer as well as the wires from the output drivers to the periphery also have shorter length.

3D Divided Bitline (3DBL) This approach is akin to the 3DWL strategy and applies partitioning to the bitlines of a sub-array (see Fig. 3.1). The bitline length in the sub-array and the number of the pass transistors connected to a single bitline are reduced. In the 3DBL approach, the sense amplifiers can either be duplicated across different device layers or shared among the partitioned sub-arrays in different layers. The former approach is more suitable for reducing access time while the latter is preferred for reducing number of transistors and leakage. In the latter approach, the sharing increases complexity of multiplexing of bitlines and reduces performance as compared to the former. Similar to 3DWL, the length of the global lines are reduced in this scheme.

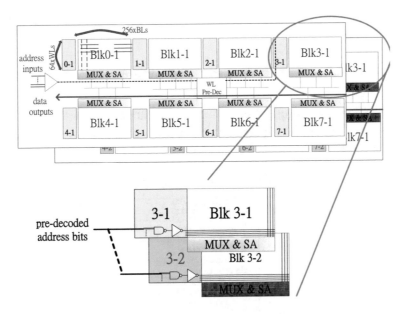

Figure 3.1: Cache with 3D divided bitline partitioning and mapped into two active device layers [40].

3D Cache Delay-Energy Estimator (3DCACTI)

In order to explore the 3D cache design space, a 3D cache delay-energy estimation tool called 3DCacti was developed [40]. The tool was built on top of the Cacti 3.0 2D cache tool (Cache Access and Cycle Time) [41]. 3DCacti searches for the optimized configuration that explores the best delay, power, and area efficiency trade-off, according to the cost function for a given number of 3D device layers.

In the original Cacti tool [41], several configuration parameters (see Table 3.1) are used to divide a cache into sub-arrays to explore delay, energy, and area efficiency trade-offs. In 3DCacti implementation, two additional parameters, Nx and Ny, are added to model the intra-subarray 3D partitions. The additional effects of varying each parameter other than the impact on length of global routing signals are listed in Table 3.1. Note that the tag array is optimized independently of the data array and the configuration parameters for tag array: Ntwl, Ntbl, and Ntspd are not listed.

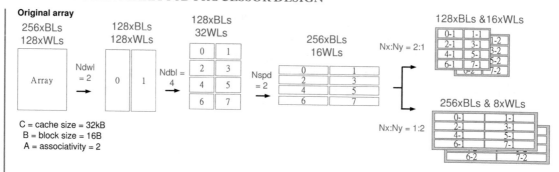

Figure 3.2: Cache with 3D divided bitline partitioning and mapped into two active device layers [5].

Figure 3.2 shows an example of how these configuration parameters used in 3DCacti affect the cache structure. The cell level partitioning approach (using MLBS) is implicitly simulated using a different cell width and height within Cacti.

Table 3.1: Design parameters for 3DCacti and their impact on cache design

Parameter	Definition	Effect on Cache Design
Ndbl	the number of cuts on a cache to divide bitlines	1. the bitline length in each sub-array 2. the number of sense amplifiers 3. the size of wordline driver 4. the decoder complexity 5. the multiplexors complexity in data output path
Ndwl	the number of cuts on a cache to divide wordlines	1. the wordline length in each sub-array 2. the number of wordline drivers 3. the decoder complexity
Nspd	the number of sets connected to a wordline	1. the wordline length in each sub-array 2. the size of wordline drivers 3. the multiplexors complexity in data output path
Nx	the number of 3D partitions by dividing wordlines	1. the wordline length in each sub-array 2. the size of wordline driver
Ny	the number of 3D partitions by dividing bitlines	1. the bitline length in each sub-array 2. the complexity in multiplexors in data output path

3.1.2   DESIGN EXPLORATION USING 3DCACTI

By using 3DCacti, we can explore various 3D partitioning options of caches to understand their impact on delay and power at the very early design stage. Note that the data presented in this section is in 70 nm technology, assuming one read/write port and one bank in each cache, unless otherwise stated.

First, we explore the best configurations for various degrees of 3DWL and 3DBL in terms of delay. Figure 3.4 and Fig. 3.5 show the access delay and energy consumption per access for 4-way set associative caches of various sizes and different 3D partitioning settings. Remember that Nx (Ny) in the configuration refers to the degree of 3DWL (3DBL) partitioning. First, we

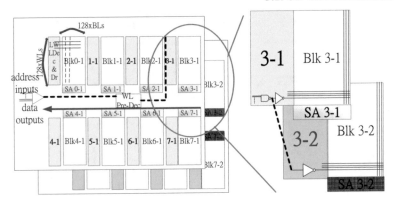

Figure 3.3: An example showing how each configuration parameter affects a cache structure. Each box is a sub-array associated with an independent decoder [40].

observe that delay reduces as the number of layers increase. From Fig. 3.6, we observe that the reduction in global wiring length of the decoder is the main reason for delay reduction benefit. We also observe that for the 2-layer case, the partitioning of a single cell using MLBS provides delay reduction benefits similar to the best intra-subarray partitioning technique as compared to the 2D design.

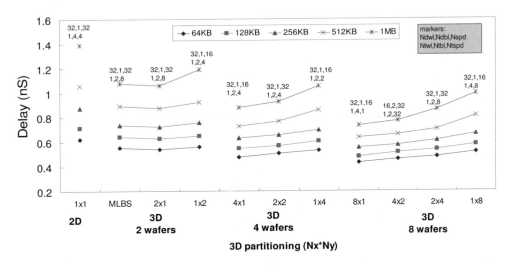

Figure 3.4: Access time for different partitioning [40]. Data of caches of associativity=4 are shown.

Another general trend observed for all cache sizes is that partitioning more aggressively using 3DWL results in faster access time. For example, in the 4-layer case, the configuration 4x1 (folding wordline into four layers) has an access time which is 16.3% less than that of the 1x4

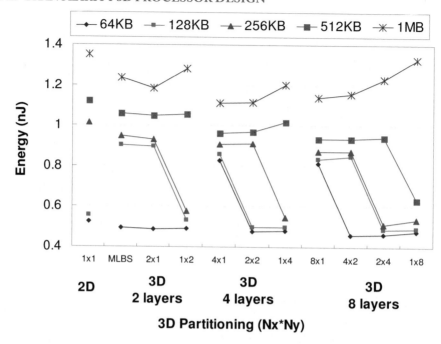

Figure 3.5: Energy for different partitioning when setting the weightage of delay higher [40]. Data of caches of associativity=4 are shown.

(folding bitline into four layers) configuration for a 1MB cache. We observed that the benefits from more aggressive 3DWL stem from the longer length of the global wires in the X direction as compared to the Y direction before 3D partitioning is performed. The preference for shorter bitlines for delay minimization in each of the sub-arrays and the resulting wider sub-arrays in optimal 2D configuration is the reason for the difference in wire lengths along the two directions. For example, in Fig. 3.8(a), the best sub-array configuration for the 1MB cache in 2D design results in the global wire length in the X direction being longer. Consequently, when wordlines are divided along the third dimension, more significant reduction in critical global wiring lengths can be achieved. Note that because 3DCacti is exploring partitioning across the dimensions simultaneously, some configurations can result in 2D configurations that have wirelengths greater in the Y directions (see Fig. 3.8(c)) as in the 1MB cache 1x2 configuration for two layers. The 3DBL helps in reducing the global wire length delays by reducing the Y direction length. However, it is still not as effective as the corresponding 2x1 configuration as both the bitline delays in the core and the routing delays are larger (see Fig. 3.6 and Fig. 3.7). These trends are difficult to analyze without the help of a tool to partition across multiple dimensions simultaneously. The energy reduction for the corresponding best delay configurations tracks the delay reduction in many cases. For example, the energy of 1MB cache increases when moving from an 8x1 configuration

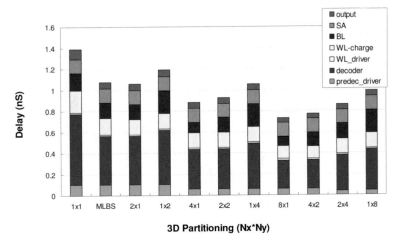

Figure 3.6: Access time breakdown of a 1MB cache corresponding to the results shown in Fig. 3.6 [40].

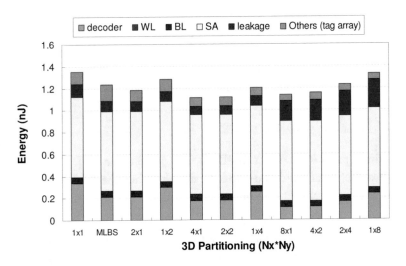

Figure 3.7: Energy breakdown of a 1MB cache corresponding to the results shown in Fig. 3.7 [40].

to a 1x8 configuration. In these cases, the capacitive loading that affects delay also determines the energy trends. In some cases, the energy reduces significantly when changing configurations and does not track performance behavior. For example, for the 512KB cache using 8-layers, the energy reduces when moving from 2*4 to 1x8 configuration. This stems from the difference in the number of sense amplifiers activated in these configurations, due to the different number of bitlines in each sub-array in the different configurations and the presence of the column decoders

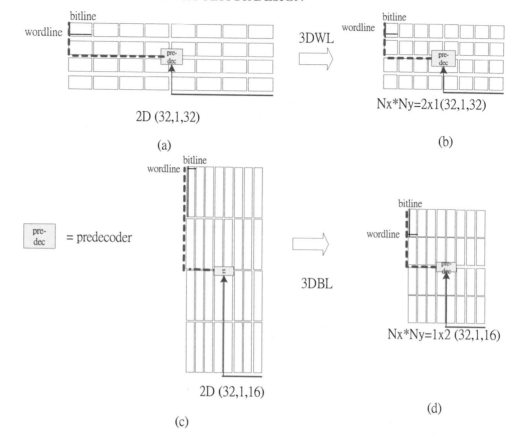

Figure 3.8: Critical paths in 3DWL and 3DBL for a 1MB cache [40]. Dashed lines represent the routing of address bits from pre-decoder to local decoder while the solid arrow lines are the routing paths from the address inputs to predecoders.

after the sense amplifiers. Specifically, the optimum (Ndwl,Ndbl,Nspd) for the 512KB case is (32,1,16) for the 2*4 case and (32,1,8) for the 1*8 configuration. Consequently, the number of sense amplifiers activated per access for 1x8 configuration is only half as much as that of the 2x4 configuration, resulting in a smaller energy.

Puttaswamy et al. provided a good study of 3D integrated SRAM components for high-performance microprocessors [34]. In this paper, they explored various design options of 3D integrated SRAM arrays with functional block partitioning. They studied two different types of SRAM arrays, which are banked SRAM arrays (e.g., caches) and multiported SRAM arrays (e.g., register files).

For the banked SRAM arrays, the paper discussed methods of bank stacking 3D and array splitting 3D. The bank stacking is straightforward as the SRAM arrays have already been parti-

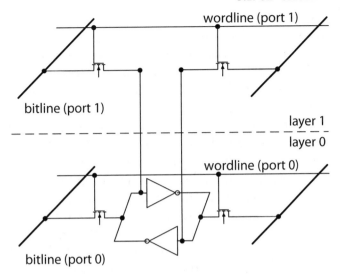

Figure 3.9: An illustration of 3D stacked two-port SRAM cell [34].

tioned into banks in the 2D case. The decision for a horizontal or vertical split largely depends on which dimension has more wire delay. The array splitting method may help to reduce the lengths of either wordlines or the bitlines. When the wordline is split, die-to-die vias are required because the row decoder needs to drive the wordlines on both of the dies, and the column select multiplexers have been split across the two dies. For the row stacked SRAM, a row decoder need to be partitioned across the two dies. If the peripheral circuitry, such as sense amplifier, is shared for the two dies, extra multiplexers may be required. In their work, latency and energy results were evaluated for caches from 16KB to 4MB. The results show that array-split configurations provide additional latency reductions as compared to the bank-stacked configurations. The 3D organization provides more benefits to the larger arrays because these structures have substantially longer global wires to route signals between the array edge and the farthest bank. For large-sized (2-4 MB) caches, the 3D organization exhibits the greatest reduction in global bank-level routing latency, since it is the dominant component of latency. On the other hand, moderate-sized (64-512 KB) cache delays are not dominated by the global routing. In these instances, a 3D organization that targets the intra-SRAM delays provides more benefit. Similar to the latency reduction, the 3D organization reduces the energy in the most wire-dominated portions of the arrays, namely the bank-level routing and the bitlines in different configurations [34].

   For the multiported SRAM, Puttaswamy et al. took the register files (RF) as an example to show different 3D stacking strategies. First, a RF can split half of the entries and stack them on the rest half. This is called register partitioning (RP). The bitline and row decoder are halved so that the access latency is reduced. Second, a bit-partitioned (BP) 3D register file stacks higher order and lower order bits of the same register across different dies. Such a strategy reduces the load on wordline. Third, the large footprint of a multiported SRAM cell provides the opportunity to allocate one or two die-to-die vias for each cell. Consequently, it is feasible that each die contains bitlines, wordlines, and access transistors for half of the ports. This strategy is called port splitting (PS), which provides benefits in reducing area footprint. Figure 3.9 gives an illustration of the 3D structure of the two-port cell. The area reduction translates into latency and energy savings. The register files with sizes ranging from 16 up to 192 entries were simulated in the work. The results show that, for 2-die stacks, the BP design provides the largest latency reduction benefit when compared to the corresponding RP and PS designs. The wordline is heavily loaded by the two access transistors per bitline column. Splitting the wordline across two dies reduces a major component of the wire latency. As the number of entries increases, benefits of latency reduction from BP designs decrease because the height of the overall structure increases, which makes the row decoder and bitline/sense-amplifier delay increasingly critical. The benefits of using the other two strategies, however, increase for the same reason. The 3D configuration that minimizes energy consumption is not necessarily the configuration that has the lowest latency. For a smaller number of entries, BP organization requires the least energy. As the number of entries increases above 64, RP organization provides the most benefits by effectively reducing bitline length [34]. When there are more than two stacked layers, the case is more complicated because of many stacking options. More details can be found in the paper.

3.2 3D PARTITIONING FOR LOGIC BLOCKS

We have discussed how to split a planar SRAM cache and integrate it with 3D stacking. Black et al. compared the performance, power consumption, and temperatures of splitting logic components of a processor and stacking them in two layers. Such a strategy is called "logic+logic" stacking. Using logic+logic stacking, a new 3D floorplan can be developed that requires only 50% of the original footprint. The goal is to reduce inter-block interconnect by stacking and reducing intra-block, or within block interconnect through block splitting [35]. Some components in pipeline, which are far from each other in the worst case data path of a 2D floorplan, can be stacked on each other in different layers so that the pipeline stages can be reduced. An example is shown in Fig. 3.10, the path between the first level data cache (D$) and the data input to the functional units (F) is drawn illustratively. The worst case path in planar processor occurs in Fig. 3.10 (a), when load data must travel from the far edge of the D$ to the farthest functional unit. Figure 3.10(b) shows that a 3D floorplan can overlap the D$ and F. In the 3D floorplan, the load data only travels to the center of the D$, at which point it is routed to the other die to the center of the functional units. As a result of stacking that same worst case path contains half as much routing distance, since the data is only traversing half of the data cache and half of the functional units [35].

Using logic+logic stacking, 25% of all pipe stages in the microarchitecture are eliminated with the new 3D floorplan simply by reducing metal runs, resulting in a 15% performance improvement [35]. Although the performance (IPC) increase causes more activity that increases the

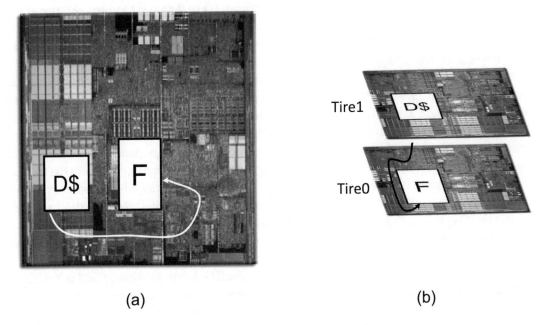

(a)

(b)

Figure 3.10: (a) Planar floorplan of a deeply pipelined microprocessor with the point register read to execute paths, (b)3D floorplan of the planar microprocessor in (a) [35].

power, the power per instruction is reduced because of the elimination of pipeline stages. The temperature increase, however, is a serious problem because of the doubling of power density in 3D stacked logic. The "worst case" shows a 26 degree increase if there were no power savings from the 3D floorplan and the stacking were to result in a 2x power density. A simple iterative process of placing blocks, observing the result is a 1.3x power density increase and 14 degree temperature increase [35].

Balaji et al. also explored design methodologies for processor components in 3D technology, in order to maximize the performance and power efficiency of these components [12]. They proposed systematic partitioning strategies for custom design of several example components: *a) instruction scheduler, b) Kogge-Stone adder,* and *c) logarithmic shifter.* In addition, they also developed a 3D ASIC design flow leveraging both widely used 2D CAD tools (Synopsys Design Compiler, Cadence Silicon Ensemble, and Synopsys Prime Time) and emerging 3D CAD tools (MIT PR3D and in-house netlist extraction and timing tool).

The custom designs are implemented with MIT 3D Magic—a layout tool customized for 3D designs—together with in-house netlist and timing extraction scripts. For the instruction scheduler, they found that the tag drive latency is a major component of the overall latency and sensitive to wire delay. Thus they proposed to either horizontally or vertically partition the tag line to reduce the tag drive latency. From the experiment result, horizontal partitioning achieves significant latency reduction (44% when moving from 2D to 2-tier 3D implementations) while vertical partitioning only achieves marginal improvement (4%). For both the KS adder and the logarithmic shifter, the intensive wiring in the design becomes the limiting factor in performance and power in advanced technology nodes. In the 3D implementation, the KS adder is partitioned horizontally and the logarithmic shifter is partitioned vertically. Significant latency reductions were observed, 20.23%–32.7% for the KS adder and 13.4% for the logarithmic shifter when the number of tiers varies. In addition to the custom design, they also experimented the proposed ASIC design flow with a range of arithmetic units. The implementation results show that significant latency reductions (9.6%–42.3%) are archived for the 3D designs generated by the proposed flow. Last but not least, it is observed that moving from one tier to two tier produces the most significant performance improvements, while this improvement saturates when more tiers are used. Architectural performance impact of the 3D components is evaluated by simulating a 3D processor running SPEC2000 benchmarks. The data path width of the processor is scaled up, since the 3D components have much lower latency than 2D components with same widths. Therefore, simulation results show an average IPC speedup of 11% for the 3D processor.

So far, the partitioning of the components in the 3D design are all specified and most partitions are limited within two layers. In order to further explore the 3D design space, Ma et al. proposed a microarchitecture-physical codesign framework to handle fine-grain 3D design [42]. First, the components in the design are modeled in multiple silicon layers. The effects of different partitioning approaches are analyzed with respect to area, timing, and power. All these approaches are considered as the potential design strategies. Note that the number of layers that a component can be partitioned into is not fixed. Especially, the single layer design of a component is also included in the total design space. In addition, the author analyzed the impact of scaling the sizes of different architectural structures. Having these design alternatives of the components, an architectural building engine is used to choose the optimized configurations among a wide range of implementations. Some heuristic methods are proposed to speed-up the convergence and reduce redundant searches on infeasible solutions. At the same time, a thermal-aware packing engine with temperature simulator tool is employed to optimize the thermal characters of the entire design so that the hotspot temperatures are below the given thermal thresholds. With these engines, the framework takes the frequency target, architectural netlist, and a pool of alternative block implementations as the inputs and finds the optimized design solution in terms of performance, temperature, or both. In the experiments, the author used a superscalar processor as an example of design space exploration using the framework. The critical components such as the issue queue and caches could be partitioned into two to four layers. The partition methods included block folding and port partitioning, as introduced before. The experimental results show a 36% performance improvement over traditional 2D and 14% over 3D with single-layer unit implementations.

We have discussed that the 3D stacking aggravates the thermal issues, and a thermal-aware placement can help alleviate the problem [35]. There are some architectural level techniques,

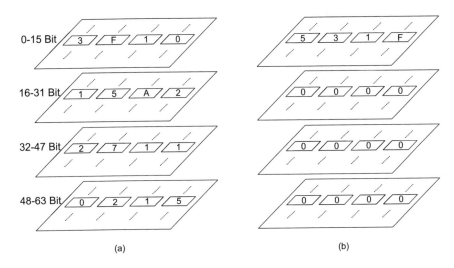

Figure 3.11: The conceptual view of a 3D stacked register file: (a) data bits on lower three layers are all zeros, (b) data in all layers are non-zeros [36].

which can be used to control the thermal hotspots. Puttaswamy et al. proposed thermal herding techniques for the fine-grain partitioned 3D microarchitecture. For different components in the processor, various techniques are introduced to reduce 3D power density and locate a majority of the power on the die (layer) closest to the heat sink.

Assume there are four 3D stacked layers and the processor is 64-bit based. Some 64-bit based components (e.g., register files, arithmetic units) are equally partitioned and 16-bit is placed in each layer, as shown in Fig. 3.11. Such a partition not only reduces the access latency, but also provides the opportunities for thermal herding. For example, in Fig. 3.11(a) the most significant 48 bits (MSBs) of the data are zero. We don't have to load/store these zero bits into register files, or process these bits in the arithmetic unit, such as the adder. If we only process the least significant 16 bits (LSBs), the power is reduced. In addition, if the LSBs are located in the layer next to the heat sink, the temperature is also decreased. For the data shown in Fig. 3.11(b), however, we have to process all bits since they are non-zeros in four layers.

For some other components, such as instruction scheduler and data caches, entries are partitioned and placed in different layers. The accesses to these components are controlled so that the entries, which are located in the layer next to the heat sink, are more frequently accessed. Consequently, the energy is herded toward the heat sink and the temperature is reduced. Extra hardware, however, is required to achieve these techniques. For example, an extra bit is induced in the register file to represent whether the MSBs are zero. This bit is propagated to other components for further controls. Compared to a conventional planar processor, the 3D processor achieves a 47.9% frequency increase which results in a 47.0% performance improvement, while simultaneously reducing total power by 20%. Without Thermal Herding techniques, the worst-case 3D temperature increases by 17 degrees. With Thermal Herding techniques, the temperature increase is only 12 degrees [36].

CHAPTER 4

Coarse-granularity 3D Processor Design

In this section, we will focus on the "memory+logic" strategy in multi-core processors. Memory of various technologies can be stacked on top of cores as caches or even on-chip main memories. Different from the research in the previous section, which focuses on optimizations in the fine-granularity (e.g., wire length reduction), the approaches of this section consider the memories as a whole structure and explore the high-level improvements, such as access interfaces, replacement policies, etc.

4.1   3D CACHES STACKING

Black et al. proposed to stack SRAM/DRAM L2 caches on cores based on Intel's $Core^{TM}$ 2 Duo model. It was assumed that the 4M L2 caches occupy 50% area in a planar two-core processor. They proposed three different stacking options without increasing the footprint: (1) stacking extra 8M L2 caches on the original processor; (2) removing 4M SRAM caches and stacking a 32M DRAM caches on logic (the 2MB DRAM tag is placed in the same layer of logic); (3) removing 4M SRAM caches and stacking 64M DRAM caches (the 4MB DRAM tag is placed in the same layer of logic). These three options are shown in Fig. 4.1 together with the 2D baseline. The results showed that stacking DRAM caches could significantly reduce the average memory access time and reduce the off-chip bandwidth for some benchmarks with a working set larger than 4MB. In addition, the off-chip bus power was greatly reduced due to the reduction of off-chip memory access. Since the DRAM has less power than SRAM of the same area, stacking DRAM cache also reduces the power compared to the case of stacking SRAM caches. The thermal simulations also showed that the peak temperature was increased by less than three degrees with the 64MB DRAM caches stacked. It meant that the thermal impact of stacking DRAM was not significant [35].

While prior approaches have shown the performance benefits of using 3D stacked DRAM as a large last-level cache (LLC), Loh proposed new cache management policies for further improvements leveraging the hardware organization of DRAM architectures [30]. The management scheme organizes the ways a cache set into multiple FIFOs (queues). As shown in Fig. 4.2, each entry of the queue represents a cache line, and all entries of queues and "LRU" cache cover a set of cache line. The first level queues and second level queues are similar except that, in multi-core processors, there is one first level of queue for each thread, but the second level queue is shared,

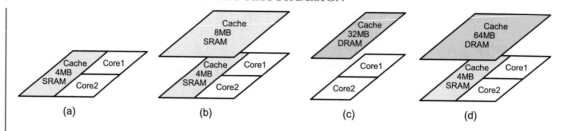

Figure 4.1: Memory stacked options: (a) 4MB baseline; (b) 8MB stacked for a total of 12MB; (c) 32MB of stacked DRAM with no SRAM; (d) 64MB of stacked DRAM [35].

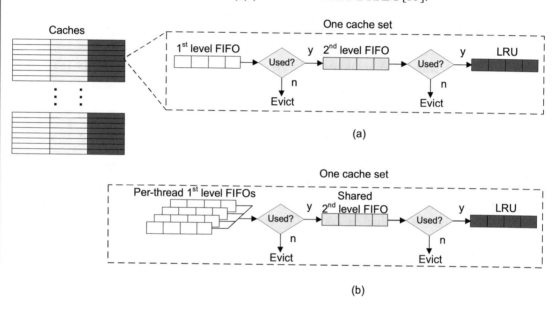

Figure 4.2: Structure of the multi-queue cache management scheme for (a) a single core and (b) multiple cores [30].

as shown in Fig. 4.2(b). The data loaded to LLC is first inserted into these queues before being placed into cache ways following "LRU" policies. A "u-bit" is employed to represent whether the cache line is re-used (hit) during its stay in these queues. When cache lines are evicted from queues, they are moved to LRU-based cache ways only if they have been re-used. Otherwise, these cache lines are evicted from queues directly. Under some cache behaviors, these queues can effectively prevent useful data from being evicted by inserted data, which are not reused.

The first level of queue is core-based so that the cache requests of different cores are placed into separated queues. Consequently, the cache replacements of different cores are isolated from each other in first-level queues. Such a scheme can help to improve the utilization efficiency of caches because a core with a high access rate can quickly evict the cache lines of another core from the shared cache without the isolation. This scheme raises another issue of how to allocate the size of the first-level queue for each core. In order to solve this problem, the paper proposed an adaptive multi-queue (AMQ) method, which leveraged the set-dueling principal [30] to dynamically allocate the size of each first-level queue. There are several pre-defined configurations of queue sizes, and the AMQ can dynamically decide which configurations should be used according to the real cache access pattern. A stability control mechanism was used to avoid unstable and rapidly switching across many different configurations of first-level queues. The results showed that the AMQ management policy can improve the performance by 27.6% over the baseline of simply using DRAM as LLC. This method, however, incurs extra overhead of managing multiple queues. For example, it requires a head pointer for each queue. In addition, it may not work well when the cache associativity is not large enough (64 way was used in the paper) because the queue size needs to be large enough to record the re-use of a cache line. Since the first level queue is core-based, it has a scalability limitation for the same reason.

4.2 3D MAIN MEMORY STACKING

In addition to the caches stacking, many studies have been exploring the potential of building main memory on-chip with the help of 3D integration [31–33, 43–45].

The most straightforward way is to implement a vertical bus across the layers of stacked DRAM layers to the processor cores [43–45]. The topology and overall architecture of the 3D stacked memory is kept the same as in the traditional off-chip memory. Previous work shows that, simply placing the DRAM provides a 34.7% increase in performance on memory-intensive workloads from SPEC2006 [31]. The benefits come from reducing wire delay between the memory controller and main memory, and running the memory controller at a faster speed.

Due to the advantages of dense TSV, the interface between cores and stacked memories can be implemented as wider buses, compared to that of 2D off-chip memory. After increasing the width of bus from 64-bit (2D FSB width) to 64-byte TSV-based 3D buses, the performance improvement over 2D case is increased to 71.8% [31]. Increasing the bus width does not fully exploit 3D stacking because the memory organization is still inherently two-dimensional. Tezzaron Corporation has announced "true" 3D DRAMs where the individual bitcell arrays are stacked in a 3D fashion [46]. Different from the prior approaches, these true 3D DRAMs isolate the DRAM cells from peripheral circuitry. The DRAM bitcells implemented in a NMOS technology is optimized for density, whereas the peripheral circuitry implemented on a CMOS layer optimized for speed. The combination of reducing bitline capacitance and using high-speed logic provides significant improvement in memory access time. Consequently, the true 3D structures result in a 116.8% performance improvement over the 2D case [31].

The high bandwidth of 3D stacking also makes it a viable option to increase the number of ranks and memory controller interfaces (MC) of 3D stacked memories. Note that it is not practical in off-chip due to pin and chip-count limitations [31]. Loh modified the banking of L2 caches to route each set of L2 banks to one and only one MC so that the interleaving of L2

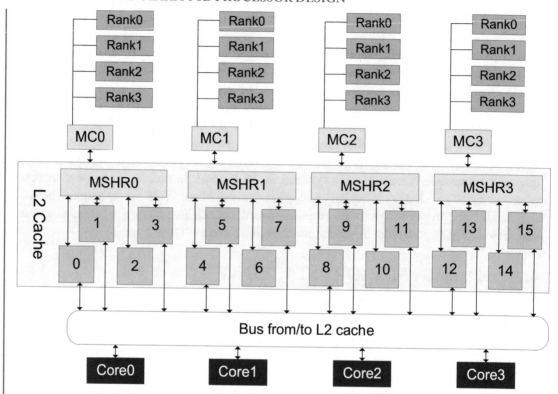

Figure 4.3: A structure view of 4 MSHR banks, 4 MCs, 16 Ranks [31].

cache occurs on 4KByte boundaries rather than 64 bytes. Such a structure makes L2 cache banks "aligned" to their own MSHR and MC so that the communication is reduced when we have multiple ranks and MCs. Figure 4.3 shows the structure of the 4-core processor with the aligned L2 cache banks and corresponding MSHRs and MCs. The results show that we can have another $1.75X$ speedup in addition to that provided by the true 3D structure.

Loh's work also pointed out that the significant increase in memory system performance makes the L2 miss handling architecture (MHA) a new bottleneck. The experiments showed that simply increasing the capacity of MSHR cannot improve the performance consistently. Consequently, a novel data structure called the Vector Bloom Filter with dynamic MSHR capacity tuning was proposed to achieve a scalable MSHR. The VBF-based MSHR is effectively a hash-table with linear probing, which speeds up the searching of MSHR. The experimental results show that the VBF-based dynamic MSHR can provide a robust, scalable, high-performance L2 MHA for 3D-stacked memory architecture. Overall, a 23.0% improvement is observed on memory-intensive workloads of SPEC2006 over the baseline L2 MSHR architecture for the dual-MC (quad-MC) configuration [31].

Kgil et al. studied the advantages of 3D stacked main memory, with respect to the energy efficiency of processors [43]. In their work, the structure of the DRAM-based main memory was not changed and was stacked directly on the processor. The bus width of the 3D main memory was assumed to be up to 1024-bit. With the large bus width, one interesting observation was that the 3D main memory could achieve a similar bandwidth as the L2 cache. Consequently, the paper pointed out that the L2 cache was no longer necessary and the space could be saved to insert more processing cores. With more processing cores, the frequency of the cores could be low down without degrading computing bandwidth. The energy efficiency could be increased because the power consumption was reduced as the frequency decreased, especially for applications with high thread level parallelism. The paper provided comprehensive comparisons among different configurations of 2D and 3D processors with similar die areas, with respect to processing bandwidth and power consumption.

Woo et al. further explored the high bandwidth of 3D stacked main memory by modifying the structure of the L2 cache and the corresponding interface between cache and 3D stacked main memory [32]. The paper first revisited the impact of cache line size on cache miss rate when there was no constraint to the bandwidth of main memory. The experimental results show an interesting conclusion that most modern applications will benefit from a smaller L1 cache line size, but a larger cache line is found to be helpful for a much larger L2 cache. Especially, the maximum line size, an entire page (4KB), is found to be very effective in a large cache. Then, the paper proposed a technique named "SMART-3D" to employ the benefits of large cache line size. The cache line of L2 cache was still kept as 64B, and read/write from L1 cache to L2 cache was operated with the traditional 64B bus structures. The bus between L2 cache and main memory, however, was designed to be 4KB, and 64 cache lines could be filled with data from main memory at the same time. In order to achieve a high parallel data filling, the L2 cache was divided into 64 subarrays, and one cache line from each subarray was written in parallel. Two cache eviction policies were proposed so that either one or 64 cache lines could be evicted on demand. Besides the modification to L2 cache, the 3D stacked DRAM was carefully re-designed since a large number (32K) of TSVs were required in SMART-3D. The paper also noticed the potential exacerbation of the false sharing problem caused by SMART-3D. The cache coherence policy was modified so that either one or 64 cache lines are fetched for different cases.

The experimental results show that the performance is improved with the help of SMART-3D. For the single core, the average speedup of SMART-3D is 2.14x over 2D case for memory-intensive benchmarks from SPEC2006, which is much higher than that of base 3D stacked DRAM. In addition, using 1MB L2 and 3D stacked DRAM with SMART-3D achieves 2.31x speedup for a two-core processor, whereas a 4M L2 cache without SMART-3D only achieves 1.97x over the 2D case. Furthermore, the analysis shows that SMART-3D can even lower the energy consumption in the L2 cache and 3D DRAM because it reduces the total number of row buffer misses.

It is known that the periodical refresh is required for DRAMs to maintain the information stored in them. Since the read access is self-destructive, the refresh in a DRAM row involves reading the stored data in each cell and immediately rewriting back to the same cell. The refresh incurs considerable power and bandwidth overhead. 3D integration is a promising technique that benefits DRAM design particularly from capacity and performance perspectives. Nevertheless, 3D stacked DRAMs potentially further exacerbates power and bandwidth overhead incurred by the refresh process. Ghosh et al. proposed "Smart Refresh" [33], a low-cost technique implemented in the memory controller to eliminate the unnecessary periodic refresh processes and mitigate the energy consumption overhead in DRAMs. By employing a time-out counter in the memory controller, for each memory row of a DRAM module, the DRAM row that was recently accessed will not be refreshed during the next periodic refresh operation. The simulation results show that the proposed technique can reduce 53% of refresh operations on average with a 2GB DRAM, and achieves 52.6% energy saving for refresh operations and 12.13% overall energy saving on average. An average 9.3% energy saving can be achieved for a 64MB 3D DRAM with 64ms refresh rate, and 6.8% energy saving can be achieved for the same DRAM capacity with 32ms refresh rate.

Chen et al. developed an architecture-level modeling tool, CACTI-3DD [47], which estimates timing, area, and power of the 3D die-stacked off-chip DRAM main memory. This tool is designed based on the DRAM main memory model of CACTI [48]; it introduces TSV models, improves the modeling of 2D off-chip DRAM main memory, and includes 3D integration modeling on top of the original CACTI memory models. CACTI-3DD enables the analysis of a full spectrum of 3D DRAM design from coarse-grained rank-level 3D stacking to bank-level 3D stacking. It allows memory designers to perform in-depth studies of 3D die-stacked DRAM main memory, in terms of architecture-level tradeoffs of timing, area, and power. Their study also demonstrated the usage of the proposed tool by re-architecting the DRAM dies at a fine granularity under the guidance of modeling results. The redesigned 3D-stacked DRAM main memory can achieve significant timing and power improvements compared with coarse-grained baseline.

4.3   3D ON-CHIP STACKED MEMORY: CACHE OR MAIN MEMORY?

3D/2.5D integration technology allows computer designers to integrate gigabytes of DRAM into the processor package. Such on-package DRAMs introduce numerous attractive characteristics to the memory system, such as high memory bandwidth, fast access speed, and large on-chip memory capacity. Yet one nuisance facing system designers is how to make efficient use of them—as a large last-level cache (LLC) or a fast main memory. The two design options trade-off among access speed, design complexity, and cost. Consequently, the answer is not straightforward. In this chapter, we introduce existing designs that choose either design options, as well as a combination of the two.

4.3.1   ON-CHIP MAIN MEMORY

Dong et al. [49] observed that adopting the on-chip DRAM as LLC is less feasible than as a portion of the main memory, due to the non-trivial design efforts required to accommodate access to a large cache capacity. Commodity DRAM dies have been optimized for cost and do not employ specialized tag arrays that automatically determine a cache hit/miss and forward the request to the corresponding data array. Because the size of a tag array can be a hundred megabytes or more with a multi-gigabyte cache, storing tags on the processor die requires an impractically large tag space. The only alternative is to place the LLC tags inside the on-chip DRAM. However, doing so will require two DRAM accesses upon each cache hit: one looking up the tag and the other returning the data. Each cache hit will take approximately $2\times$ of the time for a single access to the on-package DRAM. Their experimental results showed that there is almost no benefit to enlarge the LLC capacity in terms of the cache miss rate. While accessing the LLC and the main memory in parallel can help hide the long LLC access latency, there is not sufficient off-chip bandwidth to access the off-chip memory speculatively and simultaneously with every reference to an on-chip cache. Furthermore, off-chip references consume significantly more power and should generally be avoided when possible.

To avoid such issues, Dong et al. proposed a heterogeneous main memory architecture, which leverages the on-chip 3D-stacked DRAM as a fast portion of the main memory. In addition, the main memory adopts four DDR3 channels connected to traditional off-chip DRAM Dual In-line Memory Modules (DIMMs) to extend the total memory capacity. Figure 4.4 illustrates an overview of their heterogeneous main memory architecture. DRAM dies are placed beside the processor die using 2.5D integration technology. The flip-chip system-in-package (SiP) can provide a die-to-die bandwidth of at least 2 Tbps [50]. To reduce the memory access latency, the on-chip DRAM is slightly modified based on commodity DRAM products to implement a many-bank structure. Doing so further increased the signal I/O speed by taking the advantage of the high-speed on-chip interconnects. The proposed design did not employ a custom tag part and adopted only a single on-chip DRAM design in order to reduce design cost and maximize the effective capacity of the on-package DRAM. The on-chip memory controller is connected to off-chip DIMMs through the conventional 64-bit DDRx bus and to the on-chip memory through a customized memory bus. MSBs of physical memory addresses are used to decode the target location. For example, if 1GB of 32-bit memory space is on-package, $Addr[31..30] = 00$ is mapped to on-chip memory while $Addr[31..30] = 01, 10, 11$ is mapped to off-chip DIMMs.

Their experiments with all ten workloads in NAS Parallel Benchmark Suite 3.3 [51] showed directly mapping 1GB on-chip DRAM resources into the main memory space can always achieve better performance than using the on-chip DRAM resources to add a new L4, when application memory footprints fit into the on-chip memory. However, with workloads that have a much larger working set, the performance improvement achieved by such static mapping was trivial. For example, the performance improvement of DC.B is only 16% and that of FT.C is only 20.7%. Both of them are less than the ones achieved by using on-chip DRAM as LLC.

To solve this issue, they proposed to add data migration functionality into the memory controller so that frequently used data can remain on-chip with a higher probability. Compared with other works on data migration [52–56]: (1) their data migration was implemented by introducing another layer of address translation; (2) depending on the data granularity, they proposed either a pure-hardware implementation or an OS-assisted implementation; and (3) a novel migration algorithm was used to hide the data migration overhead. They used the term macro page as the data migration granularity, and the macro page size can be much larger than the typical 4KB page size used in most operating systems.

In particular, their data migration algorithm is based on the hottest-coldest swapping mechanism, which first monitors the LRU (least recently used) on-chip macro page (the coldest) and the MRU (most recently used) off-chip macro page (the hottest) during the latest period of execution and then triggers the memory migration if the off-chip MRU page is accessed more frequently than the on-chip LRU page after each monitoring epoch. The migration algorithm can be implemented in either a pure-hardware scheme or OS-assisted manner depending on the migration granularity. The pure-hardware solution is preferred when the macro page size is relatively large so that the scale of macro page count is controllable within certain hardware resources. If finer granularity of data migration is required, the number of macro pages becomes too large for hardware to handle and OS-assisted scheme is used to track the access information of each macro page. Basically, the functionality of the data migration is achieved by keeping an extra layer of address translation that maps the physical address to the actual machine address. The pure-hardware scheme keeps the translation table in hardware while the OS-assisted scheme keeps it in software.

Their evaluation results demonstrate how the heterogeneous main memory can use the on-package memory efficiently and achieve the effectiveness of 83% on average.

To manage such a space and move frequently accessed data to fast regions, we propose two integrated memory controller schemes: a first technique handles everything in hardware and our second scheme takes assistance from the operating system.

4.3.2   3D-STACKED LLC

Loh and Hill [57] argue that an intelligent tag design can enable the use of on-chip DRAM as a large-capacity LLC by effectively reducing the tag overhead. They performed a comprehensive analysis on the challenges and inefficiencies found in prior work to implement large caches: combining tags and data in the DRAM will cause the aforementioned performance and area challenges; managing the data with large cache lines can lead to fragmentation, false-sharing, and contention among multiple banks; sub-blocked caches still require a large tag storage space.

They proposed a new DRAM cache architecture, which supports conventional cache block sizes (e.g., 64-byte) at low performance overhead. While they adopted a combined tags and data in the DRAM, they employed two sophisticated mechanisms to accelerate accesses to the DRAM cache (shown in Fig. 4.5). First, they adopted a compound-access scheduling, which improves the cache hit latency with a simple modification of the memory controller's scheduling algorithm. In

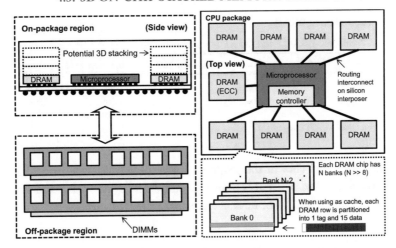

Figure 4.4: The conceptual view of the System-in-Package (SiP) solution with one microprocessor die and nine DRAM dies connecting off-package DIMMs (one on-package DRAM die is for ECC).

their design, a single physical DRAM row holds both tags and data, as illustrated in Fig. 4.5(a). Therefore, the memory controller can access the data and the corresponding tag with a single compound operation. During access to the row, the memory controller prevents subsequent memory requests from closing the row by reserving the row buffer. As a result, any updates to the data and the corresponding tag will also hit in the row buffer. Second, they developed a MissMap mechanism to avoid the DRAM cache access on a cache miss, by bypassing the DRAM cache and directing the request immediately to the off-chip main memory. Their solution is to employ a MissMap data structure (Fig. 4.5(b)) which tracks the cache lines belonging to the same page stored in the DRAM cache; each MissMap entry has a bit vector, each associated with a cache line. A zero-bit or no corresponding entry in the MissMap indicates a DRAM cache miss. In this case, the request will be directly issued to the off-chip main memory.

Their experimental results showed that the compound-access scheduling improves the performance of a DRAM L4 cache by half and provides 92.9% of the performance delivered by the ideal SRAM-tag configuration compared to having no DRAM L4 cache. With MissMap, their design offers 97.1% of the performance of the ideal SRAM-tag configuration.

Qureshi and Loh [58] further improved the DRAM cache design with a latency-optimized cache architecture that is different from conventional cache organizations. They observed that Loh and Hill's design can substantially increase the DRAM cache access latency by serializing the tag accesses and the accesses to the MissMap. Their design leverages DRAM bursts which do not exist in conventional SRAM caches by streaming tag and data together in a single burst. Their design constructs the DRAM cache as direct-mapped cache, and tightly couples the tag and data into one unit to prevent the tag serialization penalty. Doing this effectively eliminated the

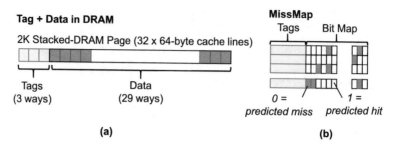

Figure 4.5: The conceptual view of the proposed stacked DRAM cache management mechanisms. (a) A single physical DRAM row holds both tags and data; (b) MissMap structure used to predict whether a memory request will hit or miss in the stacked DRAM cache.

delay caused by tag serialization and improved performance by sacrificing DRAM cache hit rate. To address the performance penalty introduced by MissMap accesses, they proposed a memory access predictor, which requires only 96-byte storage per core and provides a performance of 98% of a perfect predictor. Their evaluation across various benchmarks in SPEC2006 suite showed that the proposed design outperforms Loh and Hill's work by 24% and the ideal SRAM tag design by 8.9%.

4.3.3   DYNAMIC APPROACH

As shown in above sections, neither of the two design options appears to be substantially more efficient than the other. One trade-off between the two is main memory capacity and design complexity: the on-chip main memory extends the main memory capacity, while requiring sophisticated OS modifications to support page migration between the on-chip and the off-chip memory regions; the 3D-stacked LLC eliminates the need for OS modifications, but loses the benefit of extending the physical address space of the main memory.

Chou et al. [59] found a sweet spot in between. They proposed an on-chip memory design, namely CAMEO, which is managed as an LLC yet exposed to the OS as a portion of the physical address space. Figure 4.6 illustrates a high-level view of the proposed CAMEO design compared to previous studies. With CAMEO, the on-chip, 3D-stacked DRAM is visible to the OS and managed at the cache line granularity (referred to as memory lines in the rest of this chapter). The key to exposing the stacked DRAM to the OS is the notion of Congruence Group, which determines the memory lines that can be mapped to a given location in stacked DRAM. CAMEO retains the most recently accessed memory lines. When a miss happens at the stacked memory, CAMEO swaps the data fetched from the off-chip memory with existing memory line in the stacked DRAM. CAMEO restricts that memory lines in stacked DRAM can only be swapped with an off-chip memory line from the same Congruence Group. The number of Congruence Groups is equal to the number of lines in stacked memory.

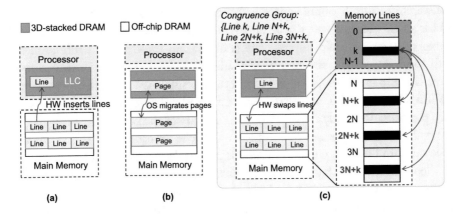

Figure 4.6: The proposed CAMEO architecture (c) compared with prior designs of 3D-stacked DRAM as LLC (a) and a portion of main memory (b) [59].

Such swapping sustains the latency and the bandwidth benefits of the stacked memory. However, it may change the physical location of a memory line without the notice of the OS. To address this issue, the proposed design employs a line location table (LLT) to track the physical location of all memory lines. CAMEO employs two methods to reduce the storage overhead yet sustain low memory access latency. First, CAMEO stores the corresponding LLT entries in stacked DRAM, residing in the same rows of these memory lines. Second, CAMEO adopts a Line Location Predictor (LLP), a hardware structure less than 1KB per core, to predict whether a memory line is exclusively stored in off-chip DRAMs and identify the potential physical addresses of the off-chip memory lines. With the two method, CAMEO can access a memory line with the latency of one memory reference, regardless of the location of the memory line. If a memory line is predicted to be off-chip, CAMEO will fetch the predicted physical address in parallel with LLT accesses.

Experiments conducted across various capacity- and latency-limited benchmarks show that CAMEO improves system performance by 69%, while employing the stacked DRAM purely as LLC and main memory only provide performance improvements of 51% and 44%, respectively. Furthermore, CAMEO achieves 98% the performance of an idealized system which employs 1GB stacked DRAM both as an LLC and main memory capacity extension.

4.4   PICOSERVER

In this section, PicoServer [20], an architecture to reduce power and energy consumption using 3D stacking technology, is introduced.

The basic idea is to stack on-chip DRAM main memory instead of using stacked memory as a larger L2 cache. The on-chip DRAM is connected to the L1 caches of each core through shared bus architecture. It offers wide low-latency buses to the processor cores and eliminates the need for an L2 cache, whose silicon area is allocated to accommodate more cores. Increasing the number of cores can help improve the computation throughput, while each core can run at a much lower frequency, and therefore result in an energy-efficient many core design. The PicoServer is a chip multiprocessor, consists of several single issue in-order processors. Each core runs at 500MHz, containing an instruction cache and a data cache, which uses a MESI cache coherence protocol. The study showed that the majority of the bus traffic is caused by cache miss traffic instead of cache coherence due to the small cache for each core. This is one motivation to stack large on-chip DRAM, which is hundreds of megabytes, using 3D technology.

In PicoServer, a wide shared bus architecture is adopted to provide high memory bandwidth and to fully take advantage of 3D stacking. A design space is explored by running simulations varying the bus width on a single shared bus, which ranges from 128 bits to 2048 bits. The impact of bus width on the PicoServer is determined by measuring the network performance. The results showed that a relatively wide data bus is needed to achieve better performance and satisfy the outstanding cache miss requests. The bus traffic increase caused by narrowed bus width will result in latency increase. Wide bus widths can speedup DMA transfer since more data can be copied in one transaction. The simulation also shows that a 1024-bit bus width is reasonable for different configurations including 4, 8, and 12 multiprocessors due to performance saturation at this point.

The stacked on-chip DRAM contains four layers in order to obtain a total size of 256 MB, which may be enough depending on the workload. More on-chip DRAM capacity can be obtained with aggressive die stacking. In order to fully take advantage of 3D stacking, it is necessary to modify the conventional DDR2 DRAM interface for PicoServer's 3D stacked on-chip DRAM. In the conventional DDR2 DRAMs, a small pin count is assumed. In addition, address multiplexing and burst mode transfer are used to compensate the limited number of pins. With 3D stacking, there is no need to address multiplexing so that the additional logic to latch and mux narrow address/data can be removed. In servers with large network pipes such as PicoServer, one common problem is how to handle large amount of packets that arrive at each second. Interrupt coalescing, a method to coalesce non-critical events to reduce the number of interrupts, is one solution to solve this problem. However, even with this technique, the number of interrupts received by a low frequency processor in PicoServer is huge. To address this issue, multiple network interface controllers (NICs) with their interrupt lines are routed to a different processor. One NIC is inserted for two processors to fully utilize each processor. Such NIC should have multiple interface IP addresses or an intelligent method to load balancing packets to multiple processors. In addition, it needs to keep track of network protocol states at the session level.

Thermal evaluation showed that the maximum junction temperature increase is about 5 to 10°C in 5-layer PicoServer architecture. This comes from power and energy reduction caused by core clock frequency reduction and improvement on high network bandwidth.

CHAPTER 5

3D GPU Architecture

Graphics processing units (GPUs) are an attractive solution for both graphics and general purpose workloads, which demand high computational performance. 3D integration is an attractive technology in developing high-performance, power-efficient GPU systems. Recently, 3D integration has been explored by both academia and industry as a promising solution to improve GPU performance and address increasingly critical GPU power issues. In this section, we will introduce recent research and implementation efforts in 3D GPU system design.

5.1   3D-STACKED GPU MEMORY

3D-stacked GPU memory can provide an order of magnitude more connections between the GPU and the graphics memory, offering much higher memory bandwidth than traditional GDDR memories. Keckler et al. [60] envisioned that 3D stacking is a promising solution to creating dense DRAM cubes or combining GPU and DRAM stacking with interposers to accommodate high memory bandwidth requirement of parallel computing applications running on GPUs. Zou et al. [61] demonstrated the benefits of leveraging 3D integration to develop in-package nonvolatile memories in CPU/GPU heterogeneous systems. Lee et al. [62] demonstrated that 3D integration technologies can significantly reduce the power dissipation of buses in GPUs; the authors developed a GPU design that can achieve up to 21.5% power reduction compared with the 2D baseline.

GPU exploits extreme multithreading to target high-throughput [63], [64]. For example, AMD Radeon™HD 7970 employs 20,480 threads interleaved across 32 compute units [63]. To accommodate such high-throughput demands, the power consumption of GPU systems continues to increase. Zhao et al. [65] observed that up to 30% of the total GPU system power is consumed by its off-chip DRAM, i.e., graphics memory. As such, reducing the graphics memory power is critical to mitigate the power challenges facing GPU systems. To optimize GPU system energy efficiency, they proposed a "3D+2.5D" system, where the DRAM itself is 3D stacked memory with through-silicon vias (TSVs), whereas the DRAM and the GPU processor are integrated with the interposer solution (2.5D).

The proposed design explored two key power saving opportunities introduced by such in-package graphics memory. First, in-package graphics memory has a wide memory interface, which enables memory power reduction without degrading the peak memory bandwidth. The peak memory bandwidth is determined by the bus width and the frequency of the memory interface. In-package graphics memory has a bus width several times wider than conventional

GDDRs. Therefore, we can reduce the memory frequency, and still achieve equivalent or even higher peak memory bandwidth compared with conventional GDDR memories. One opportunity for memory power reduction is therefore to scale down the memory's supply voltage corresponding to the frequency reduction. Second, in-package DRAMs do not require on-die termination [66]. Therefore, we can further reduce graphics memory power by eliminating the on-die termination resistors.

Figure 5.1 illustrates the peak memory bandwidth and maximum total power consumption of 2GB graphics memory with various memory interface configurations. The graphics memories considered here are electrically similar to GDDR DRAMs except for the wider buses. The memory interface configuration is defined as a set of parameters (*bus width, frequency, Vdd*). The supply voltage (Vdd) is scaled appropriately to support the given memory interface clock frequency. DRAMs with per-channel bus widths of 16 and 32 bits are evaluated as off-chip GDDR5 memory. DRAMs with per-channel bus widths of 64 bits or wider are evaluated as in-package graphics memory. The bars in Figs. 5.1(a)–(e) show the maximum power consumption with several configurations that adopt various bus widths and clock speeds while maintaining the same peak memory bandwidth. As illustrated in Figs. 5.1(a)–(c), the memory power follows U-shaped bathtub curves at low peak memory bandwidths of 144 GB/s, 180 GB/s, and 288 GB/s. With wider memory buses, lower frequency allows us to scale down the supply voltage which directly results in a power reduction. However, the power of I/O output drivers keeps increasing along with the increase of the bus width. When the bus width is increased to 256 bits, the I/O power component starts to dominate the total memory power and finally overwhelms the power benefits of voltage and frequency (VF) scaling. With peak memory bandwidth of 144 GB/s, 180 GB/s, and 288 GB/s, the optimal bus configuration is 128-bit wide. With a higher peak memory bandwidth of 360 GB/s and 720 GB/s (Figs. 5.1(d) and (e)), the memory power will continue to decrease even when the bus width reaches 256 bits. In this case, the optimal bus width is achieved by even wider bus configurations. In both cases, the optimal bus configurations fall in the range of in-package graphics memory with wide buses.

The proposed energy-efficient GPU design leverages such power benefits of in-package graphics memory. Figure 5.3 depicts an overview of the proposed "3D+2.5D" GPU system architecture. Different from conventional GPU systems (shown in Fig. 5.2), which employ off-chip DRAMs as graphics memory, the proposed design integrates GPU processor and DRAMs with silicon interposer technology. Multiple layers of memory cells can be stacked on top of a logic layer with TSV technology. The silicon interposer area required by integrating DRAMs can be estimated based on DRAM density. DRAM density at 50nm is $27.9\,Mb/mm^2$ [67]. The density becomes $43.6\,Mb/mm^2$ with a technology node of 40nm. In this case, we can integrate 6GB of DRAMs within $281.8\,mm^2$ of silicon interposer area by stacking the DRAMs into four layers of memory cells and one logic layer.

Thermal tolerance can be one concern of processor-memory integration. Zhao et al. studied the integration of GPU processor and DRAMs with both vertical 3D memory stacking and

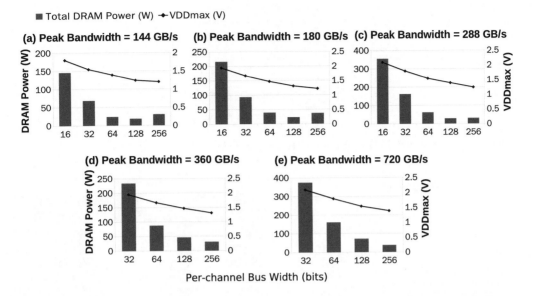

Figure 5.1: Peak memory bandwidth and maximum total DRAM power consumption of 2GB graphics memory with various interface configurations.

silicon interposer technologies. They performed thermal analysis with a GPU system configuration based on NVIDIA Quadro 6000 [64]. They computed the maximum power consumption of GPU processors and memory controllers by subtracting the DRAM power from the reported maximum power consumption of Quadro 6000 [64], resulting in 136W. The power of 6GB DRAM is calculated as 68W, based on Hynix's GDDR5 memory [68]. The areas of different GPU components are obtained from the GPU die photo, which has a $529 mm^2$ die area. They assume that the ambient temperature to be 40 °C. They used the HotSpot thermal simulation tool [69] to conduct the analysis. The maximum steady-state temperature of the GPU (without DRAMs) is 71.2 °C. With 6GB interposer-mounted DRAMs (four layers of memory cells plus one layer of logic) placed beside the GPU processor as shown in Fig. 5.3, the maximum temperature is 76.6 °C. Thus, it is feasible to employ interposer-based memory integration. Vertically stacking memories on top of the GPU incurs much greater temperature increases than a silicon interposer-based approach. By stacking the same DRAMs on top of the GPU processor, the temperature rises to 83.8 °C, a 7.2 °C increase compared to the interposer-mounted DRAMs. Moreover, the temperature rise can further increase system-wide power consumption due to the temperature dependence of leakage power. Therefore, the proposed GPU system design employs 2.5D (interposer-based) memory integration.

The memory interface with fixed bus width and frequency cannot satisfy various memory utilization requirements of various applications. Even a single application can have variable

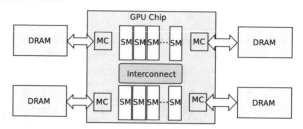

Figure 5.2: Conventional GPU system with off-chip GDDRs.

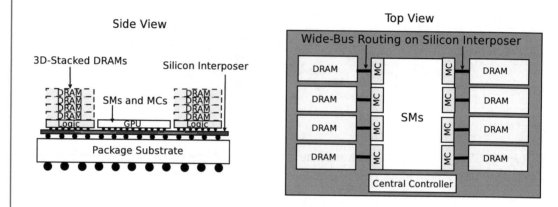

Figure 5.3: Overview of the proposed "3D+2.5D" GPU system architecture.

memory access patterns during execution. To accommodate the varying graphics memory access demands, the proposed design developed a reconfigurable memory interface, which can dynamically adapt to the demands of various applications based on dynamically observed memory access and performance information. To maximize the system energy efficiency, the proposed design configures the memory interface to minimize the DRAM power and maintain the system instructions per cycle (IPC) rate. To improve the system throughput under a given power budget, the proposed design co-optimizes the memory configuration and the GPU clock frequency by shifting power saved from the memory interface over the GPU. The proposed design employs two reconfiguration mechanisms, EOpt and PerfOpt, to optimize the system energy efficiency and system throughput, respectively. These reconfiguration mechanisms can effectively accommodate both memory-intensive and compute-intensive applications.

EOpt: The proposed design adopts a reconfigurable memory interface that can dynamically detect the various memory access patterns of the two types, and apply appropriate strategies to achieve their design goals. Both IPC and memory power will be affected by the change of memory interface during memory-intensive execution periods: decreasing the memory frequency (and consequently increasing the memory access latency) results in significant IPC degradation, even though we provide wider buses to keep the same peak memory bandwidth; the corresponding IPC typically stays much lower than that of compute-intensive periods, due to the continuous memory demands that significantly slow down the instruction execution. Therefore, they choose configurations that maintain high memory clock frequencies to minimize the IPC degradation. Given the memory frequency constraint, the bus width is then configured to minimize the memory power consumption. During compute-intensive execution periods, IPC is stable with various memory interface configurations. During these execution periods, the proposed design adopted the memory frequency and bus width configuration that minimizes the memory power.

PerfOpt explores GPU system performance (instruction throughput, i.e., the executed instructions per second) optimization under a given power budget. During compute-intensive execution periods, PerfOpt always employs the memory interface configuration that minimizes the DRAM power. Any power saved is transferred to scale up the GPU core clock frequency/supply voltage to improve the system performance. During memory-intensive periods, their design employs two strategies. First, because the memory interface configuration directly affects system performance during memory-intensive periods, they choose the memory configuration that delivers the highest system performance while staying within the system power budget. Second, sometimes an application can be relatively memory-intensive while still having significant compute needs as well. In these cases, reconfiguring the memory interface to free up more power for the GPU can still result in a net performance benefit despite the reduced raw memory performance. Based on the predicted benefit, PerfOpt will choose the better of these two strategies.

Implementation: Figure 5.4 illustrates the hardware design of the proposed reconfigurable memory interface. The design made several modifications to the interface between the GPU processor and the 3D die-stacking graphics memories, including adding a central controller, control signals to the bus drivers, and controls for dynamic VF scaling. The central controller in Fig. 5.4(a) is used to collect global information of both GPU performance and memory accesses. A vector of counters are maintained in the controller, including instruction counter, cycle counter, and memory access counters, to collect performance and memory access information from either GPU hardware performance counters or memory controllers. A threshold register vector is used to store various thresholds and initial values described in reconfiguration mechanism. The calculator module calculates the system energy efficiency based on the collected performance information and the estimated power consumption. The results are stored in result registers for comparison. Figures 5.4(b) and (c) illustrate their data bus implementation. The basic topology of a bi-directional point-to-point data bus is a set of transmission lines, with transmitter and receiving devices at both ends of each bit. Control signals of I/O drivers are connected to the central controller. These control signals switch the drivers in the transmitters on and off to change the bus width.

The experimental results show that even with a static (no reconfiguration) in-package graphics memory solution, the energy efficiency (performance per Watt) of the GPU system can be improved by up to 21%. Of course, fixing the system bandwidth to be equal to the off-chip solution does not really take advantage of the wide interface provided by the in-package memory.

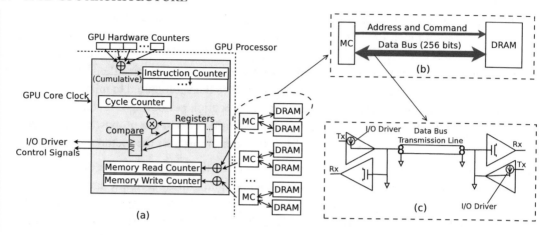

Figure 5.4: Hardware implementation of reconfigurable memory interface. (a) Central controller; (b) connection between memory controller and 3D die-stacking graphics memory (one channel); (c) reconfigurable data bus.

By increasing the memory interface clock speed to provide bandwidths of 360 GB/s and 720 GB/s (note even at these higher bandwidths, the power of in-package memory can still be lower than the off-chip GDDR5 due to the lower clock frequency), performance on the memory-intensive applications can be brought back up.

With EOpt, system power with almost all applications is reduced, and by an average of up to 12%. The overall system performance-per-Watt rate is improved for all the benchmarks. The improvement of the non-memory intensive applications (16%) is not as great as for the memory-intensive applications (44%). The reason is that the system throughput is significantly improved with memory-intensive applications, but almost stays the same with non-memory intensive applications. Overall, EOpt improves system energy efficiency of all baseline configurations, including those peak memory bandwidth configurations. Across all low- and high-intensity applications, the performance-per-Watt improves by 26% on average.

Evaluation with a variety of GPU system power budgets shows that PerfOpt can adjust the memory power consumption to fit the application memory needs, and that the saved power can be effectively redeployed to improve the GPU core performance up to 31%. For non-memory intensive applications, a higher power budget directly leads to more performance improvement. Since we always configure the memory interface to minimize the DRAM power, extra power is available to increase GPU core clock frequency. The throughput improvement of memory-intensive applications is not as significant as the non-memory intensive applications, yet PerfOpt still yields an average of 8% throughput improvement with these three most memory-intensive applications under the power budget of 220W, and more improvement with higher power budgets.

5.2   3D-STACKED GPU PROCESSOR

One aggressive approach to adopt 3D integration in GPU system design is to stack GPU caches and cores using 3D technology. For example, Maashri et al. [70] proposed a 3D GPU design with cache stacking. The work performed a comprehensive study across performance, cost, power, and thermal of 3D-stacked GPU system. The study showed that 3D-stacked GPU system can sustain low cache access latency, while increasing the cache capacity. It also showed that 3D-stacked GPU system can achieve up to 45% speedup over the 2D planar architecture without significant increase of peak temperature.

Future Products.   In-package graphics memory has been considered one of the most promising and practical solutions for energy-efficient GPU systems. Top GPU vendors, such as NVIDIA and AMD, have recently invested significant effort in investigating in-package graphics memory in their next generation products. For example, NVIDIA recently announced that their next generation Pascal GPU products will adopt 3D-stacked memory [3]. In the roadmap, NVIDIA plans to employ stacks of DRAM chips into dense modules with wide interfaces, and integrate them inside the same package as the GPU processor. Doing so will not only boost GPU system throughput and energy efficiency by allowing GPU processors to quickly access data from memory, but also allows the vendor to build more compact GPUs with much larger graphics memory capacity.

CHAPTER 6

3D Network-on-Chip

Network-on-chip (NoC) is a general purpose on-chip interconnection network architecture that is proposed to replace the traditional design-specific global on-chip wiring, by using switching fabrics or routers to connect processor cores or processing elements (PEs). Typically, the PEs communicate with each other using a packet-switched protocol, as illustrated by Fig. 6.1.

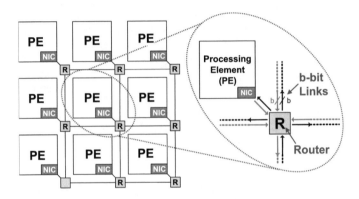

Figure 6.1: For a network-on-chip (NoC) architecture, processing elements (PEs) are connected via a packet-based network.

A typical 2D NoC consists of a number of Processing Elements (PE) arranged in a grid-like mesh structure, much like a Manhattan grid. The PEs are interconnected through an underlying packet-based network fabric. Each PE interfaces to a network router through a Network Interface Controller (NIC). Each router is, in turn, connected to four adjacent routers, one in each cardinal direction. A generic NoC router architecture is illustrated in Fig. 6.2. The router has P input and P output channels/ports. As previously mentioned, $P = 5$ in a typical 2D NoC router, giving rise to a 5×5 crossbar. The Routing Computation unit, RC, operates on the header flit (a flit is the smallest unit of flow control; one packet is composed of a number of flits) of an incoming packet and, based on the packet's destination, dictates the appropriate output Physical Channel/port (PC) and/or valid Virtual Channels (VC) within the selected output PC. The routing can be deterministic or adaptive. The Virtual channel Allocation unit (VA) arbitrates between all packets competing for access to the same output VCs and chooses a winner. The Switch Allocation unit (SA) arbitrates between all VCs requesting access to the crossbar. The winning flits can then traverse the crossbar and move on to their respective output links. Without loss of generality, all implementations in this work employ two-stage routers.

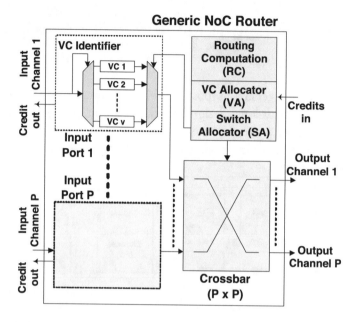

Figure 6.2: A generic NoC router architecture.

As a communication fabric, network-on-chip takes advantage of 3D integration in two ways: a) wire-length reduction which allows links to run faster and more power efficiently and b) the implementation of truly 3D network topologies which improves connectivity and reduces network diameter, and therefore promotes both performance and power.

Researchers have studied various NoC router design with 3D integration technology. For example, various design options the NoC router for 3D NoC has been investigated: 1) symmetric NoC router design with a simple extension to the 2D NoC router; 2) NoC-bus hybrid router design which leverage the inherent asymmetry in the delays in a 3D architecture between the fast vertical interconnects and the horizontal interconnects that connect neighboring cores; 3) true 3D router design with major modification as dimensionally decomposed router [71]; 4) multi-layer 3D NoC router design which partitions a single router to multiple layers to boost the performance and reduce the power consumption [72]. 3D NoC topology design was also investigated [73, 74].

Even though both 3D integrated circuits and NoCs [75, 76] are proposed as alternatives for the interconnect scaling demands, the challenges of combining both approaches to design three-dimensional NOCs have not been addressed until recently [72, 73, 77–82].

This chapter gives a brief introduction on the exploration of possible architectural designs for three-dimensional NoC architectures, and discusses the trade-offs among various design options.

Table 6.1: Area and power comparison of the crossbar switches implemented in 90 nm technology

Crossbar Type	Area	Power (500 Mhz)
5×5 Crossbar	8523 μm^2	4.21 mW
6×6 Crossbar	11579 μm^2	5.06 mW
7×7 Crossbar	17289 μm^2	9.41 mW

6.1   3D NOC ROUTER DESIGN

The NoC router design is critical for NoC architecture. Redesigning a router with the 3D technology poses interesting design challenges. Given that on-chip networks are severely constrained in terms of area and power resources, while at the same time they are expected to provide ultra-low latency, the key issue is to identify a reasonable trade-off between these contradictory design goals. In this section, we explore the extension of baseline 2D NoC router implementation into the third dimension, while considering the aforementioned constraints.

Symmetric NoC Router Design.   The natural and simplest extension to the baseline NoC router to facilitate a 3D layout is simply adding two additional physical ports to each router; one for Up and one for Down, along with the associated buffers, arbiters (VC arbiters and Switch Arbiters), and crossbar extension. We can extend a traditional NoC fabric to the third dimension by simply adding such routers at each layer (called a symmetric NoC, due to symmetry of routing in all directions). We call this architecture a 3D Symmetric NoC, since both intra- and inter-layer movement bear identical characteristics as hop-by-hop traversal. For example, moving from the bottom layer of a 4-layer chip to the top layer requires three network hops. This architecture, while simple to implement, has a few major inherent drawbacks.

- It wastes the beneficial attribute of a negligible inter-wafer distance in 3D chips (for example, the thickness of a die could be as small as 10s of μm). Since traveling in the vertical dimension is multi-hop, it takes as much time as moving within each layer. Of course, the average number of hops between a source and a destination does decrease as a result of folding a 2D design into multiple stacked layers, but inter-layer and intra-layer hops are indistinguishable. Furthermore, each flit must undergo buffering and arbitration at every hop, adding to the overall delay in moving up/down the layers.

- The addition of two extra ports necessitates a larger 7×7 crossbar. Crossbars scale upward very inefficiently, as illustrated in Table 6.1. This table includes the area and power budgets of all crossbar types investigated in this section, based on synthesized implementations in 90 nm technology. Clearly, a 7×7 crossbar incurs significant area and power overhead over all other architectures. Therefore, the 3D Symmetric NoC implementation is a somewhat naive extension to the baseline 2D network.

3D NoC-Bus Hybrid Router Design [77].   There is an inherent asymmetry in the delays in a 3D architecture between the fast vertical interconnects and the horizontal interconnects that connect neighboring cores due to differences in wire lengths (a few tens of μm in the vertical direction as compared to a few thousands μm in the horizontal direction). Consequently, a symmetric NoC architecture with multi-hop communication in the vertical (inter-layer) dimension is not desirable.

Given the very small inter-layer distance, single-hop communication is, in fact, feasible. This technique revolves around the fact that vertical distance is negligible compared to intra-layer distances; the bus can provide single-hop traversal between any two layers. This realization opens the door to a very popular shared-medium interconnect, the bus. The NoC router can be hybridized with a bus link in the vertical dimension to create a 3D NoC-Bus Hybrid structure, as shown in Fig. 6.3. This hybrid system provides both performance and area benefits. Instead of an unwieldy 7×7 crossbar, it requires a 6×6 crossbar (Fig. 6.3), since the bus adds a single additional port to the generic 2D 5×5 crossbar. The additional link forms the interface between the NoC domain and the bus (vertical) domain. The bus link has its own dedicated queue, which is controlled by a central arbiter. Flits from different layers wishing to move up/down should arbitrate for access to the shared medium.

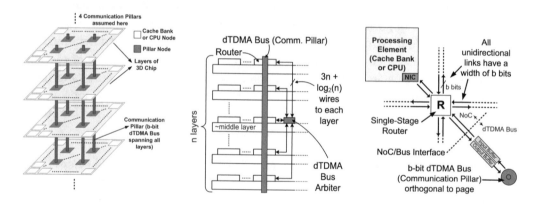

Figure 6.3: The 3D bus-hybrid NoC proposed by Li et al. [29]. (a) The vertical buses interconnecting all nodes in a cylinder; (b) the dTDMA bus and the arbiter; (c) the 3D router with the capability for vertical communication.

Despite the marked benefits over the 3D Symmetric NoC router, the bus approach also suffers from a major drawback: it does not allow concurrent communication in the third dimension. Since the bus is a shared medium, it can only be used by a single flit at any given time. This severely increases contention and blocking probability under high network load. Therefore, while single-hop vertical communication does improve performance in terms of overall latency, inter-layer bandwidth suffers. More details on the 3D NoC-Bus hybrid architecture can be found in [77].

True 3D Router Design. Moving beyond the previous options, we can envision a true 3D crossbar implementation, which enables seamless integration of the vertical links in the overall router operation. The traditional definition of a crossbar—in the context of a 2D physical layout—is a switch in which each input is connected to each output through a single connection point. However, extending this definition to a physical 3D structure would imply a switch of enormous complexity and size (given the increased numbers of input- and output-port pairs associated with the various layers). Therefore, we chose a simpler structure which can accommodate the interconnection of an input to an output port through more than one connection point. While such a configuration can be viewed as a multi-stage switching network, we still call this structure a crossbar for the sake of simplicity. The vertical links are now embedded in the crossbar and extend to all layers. This implies the use of a 5×5 crossbar, since no additional physical channels need to be dedicated for inter-layer communication.

As shown in Table 6.1, a 5×5 crossbar is significantly smaller and less power-hungry than the 6×6 crossbar of the 3D NoC-Bus Hybrid and the 7×7 crossbar of the 3D Symmetric NoC. Interconnection between the various links in a 3D crossbar would have to be provided by dedicated connection boxes at each layer. These connecting points can facilitate linkage between vertical and horizontal channels, allowing flexible flit traversal within the 3D crossbar.

The 2D crossbars of all layers are physically fused into one single three-dimensional crossbar. Multiple internal paths are present, and a traveling flit goes through a number of switching points and links between the input and output ports. Moreover, flits re-entering another layer do not go through an intermediate buffer; instead, they directly connect to the output port of the destination layer. For example, a flit can move from the western input port of layer 2 to the northern output port of layer 4 in a single hop.

However, despite this encouraging result, there is an opposite side to the coin which paints a rather bleak picture. Adding a large number of vertical links in a 3D crossbar to increase NoC connectivity results in increased path diversity. This translates into multiple possible paths between source and destination pairs. While this increased diversity may initially look like a positive attribute, it actually leads to a dramatic increase in the complexity of the central arbiter, which coordinates inter-layer communication in the 3D crossbar. The arbiter now needs to decide between a multitude of possible interconnections, and requires an excessive number of control signals to enable all these interconnections. A full crossbar with its overwhelming control and coordination complexity poses a stark contrast to this frugal and highly efficient design methodology. Moreover, the redundancy offered by the full connectivity is rarely utilized by real-world workloads, and is, in fact, design overkill [78].

Even if the arbiter functionality can be distributed to multiple smaller arbiters, then the coordination between these arbiters becomes complex and time-consuming. Alternatively, if dynamism is sacrificed in favor of static path assignments, the exploration space is still daunting in deciding how to efficiently assign those paths to each source-destination pair. Furthermore, a full 3D crossbar implies 25 (i.e., 5×5) Connection Boxes per layer. A four-layer design would, therefore, require 100 CBs! Given that each CB consists of 6 transistors, the whole crossbar structure would need 600 control signals for the pass transistors alone! Such control and wiring complexity would most certainly dominate the whole operation of the NoC router. Pre-programming static control sequences for all possible input-output combinations would result in an oversize table/index; searching through such table would incur significant delays, as well as area and power overhead. The vast number of possible connections hinders the otherwise streamlined functionality of the switch. Note that the prevailing tendency in NoC router design is to minimize operational complexity in order to facilitate very short pipeline lengths and very high frequency. A full crossbar with its overwhelming control and coordination complexity poses a stark contrast to this frugal and highly efficient design methodology. Moreover, the redundancy offered by the full connectivity is rarely utilized by real-world workloads, and is, in fact, design overkill [78].

3D Dimensionally Decomposed NoC Router Design [78].   Given the tight latency and area constraints in NoC routers, vertical (inter-layer) arbitration should be kept as simple as possible. Consequently, a true 3D router design, as described in the previous subsection, is not a realistic option. The design complexity can be reduced by using a limited amount of inter-layer links. This subsection describes a modular 3D decomposable router (called Row-Column-Vertical (RoCoVe) Router) [78].

In a typical two-dimensional NoC router, the 5×5 crossbar has five inputs/outputs correspond to the four cardinal directions and the connection from the local PE. The crossbar is the major contributor to the latency and area of a router. It has been shown [83] that through the use of a preliminary switching process known as Guided Flit Queuing, incoming traffic can be decomposed into two independent streams: (a) East-West traffic (i.e., packet movement in the X dimension), and (b) North-South traffic (i.e., packet movement in the Y dimension). Such segregation of traffic flow allows the use of smaller crossbars and the isolation of the two flows in two independent router sub-modules, which are called *Row Module* and *Column Module* [83].

With the same idea of traffic decomposition, the traffic flow in 3D NoC can be decomposed into three independent streams: with a third traffic flow in the Z dimension (i.e., inter-layer communication). An additional module is required to handle all traffic in the third dimension, and this module is called *Vertical Module*. In addition, there must be links between Vertical Module and Row/Column Modules, to allow the movement of packets from the Vertical Module to the Row Module and Column Module. Consequently, such a dimensionally decomposed approach allows for a much smaller crossbar design (4×2), resulting in a much faster and power-efficient 3D NoC router design. The architectural view of such 3D dimensionally decomposed NoC router design is shown in Fig. 6.4(b). More details can be found in [78].

Multi-layer 3D NoC Router Design [72].   All the 3D router design options discussed earlier (symmetric 3D router, 3D NoC-Bus hybrid router, true 3D router, and 3D dimensionally decomposed router) are based on the assumption that the processing element (PE) (which could be a processor core or a cache bank) itself is still a 2D design. For a fine-granularity design of

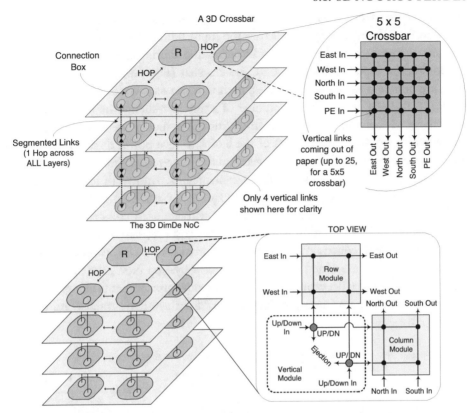

Figure 6.4: Two 3D NoC router design [84]. (a) A true 3D crossbar; (b) the dimensionally decomposed (DimDe) architecture.

3D design, one can split a PE across multiple layers. For example, 3D cache design [40] and 3D functional units [5] have been proposed before. Consequently, a PE in the NoC architecture is possible to be implemented with such fine-granularity approach. Although such a multi-layer stacking of a PE is considered aggressive in the current technology, it could be possible with monolithic 3D integration or with very small TSVs.

With such a multi-layer stacking of processing elements in the NoC architecture, it is necessary to design a multi-layer 3D router that is designed to span across multiple layers of a 3D chip. Logically, such NoC architecture with multi-layer PEs and multi-layer routers is identical to the traditional 2D NoC case with the same number of nodes albeit the smaller area of each PE and router and the shorter distance between routers. Consequently, the design of a multi-layer router requires no additional functionality as compared to a 2D router and only requires distribution of the functionality across multiple layers.

In such multi-layer router design, router components are identified as *separable* and *non-separable* modules. The *separable* modules are the input buffer, the crossbar and the inter-router links; the *non-separable* modules are the arbitration and routing logics. To decompose the separable modules, the input buffer (Fig. 6.5), the crossbar (Fig. 6.6) as well as inter-router links (Fig. 6.7), are designed as bit-slice modules, such that the data width of each component is reduced from W to W/n where n is the number of layers. Bit-slicing the input buffer reduces the length of the word-line and saves power. In addition, bit-slicing the input buffer allows for selectively switching off any unused slices of the buffer to save power during run time. The same techniques can also be applied for the crossbar and the inter-router links. As a result of 3D partitioning, the area of the router reduces and the available link bandwidth for each router increases. This excess bandwidth is leveraged to construct express physical links in this work and is shown to significantly improve performance. In addition, with reduced crossbar size and link length, the latencies of both crossbar and link stages decrease and they can be combined into one stage without violating timing constraints. The proposed architecture with express links perform best among the architectures examined (2D NoC, 3D hop-by-hop NoC, the proposed architecture without express links and with express links). The latency improvements are up to 51% and 38%, and the power improvements are up to 42% and 67% for synthetic traffics and real workloads, respectively [72].

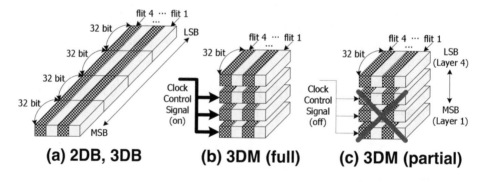

Figure 6.5: Decomposition of the input buffer [85]. (a) The 2D/3D baseline input buffer; (b) the 3D decomposed input buffer, and (c) the decomposed buffer with unused portion powered off.

6.2   3D NOC TOPOLOGY DESIGN

All the router designs discussed in previous subsections are based on the mesh-based NoC topology. There exist various NoC topologies, such as concentrated mesh or flattened butterfly topology, all of which have advantages and disadvantages. By employing different topologies rather than the mesh topology, the router designs discussed above could also have different variants. For example, in 2D concentrated mesh topology, the router itself has a radix of 8 (i.e., an 8-port

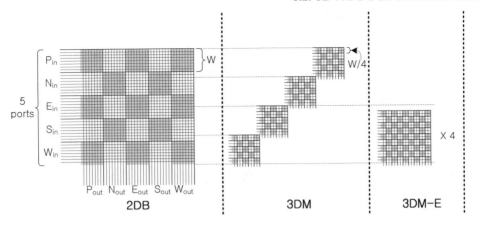

Figure 6.6: Decomposition of the crossbar [85]: the 2D baseline crossbar (2DB), the 3D decomposed crossbar (3DM), and the 3D decomposed crossbar with support for express channels (3DM-E), which sightly increases the crossbar size.

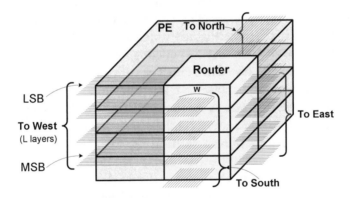

Figure 6.7: Decomposition of the inter-router links [85].

router, with four to local PEs and the others to four cardinal directions). With such topology, the 3D NoC-bus hybrid approach would result in a 9-port router design. Such high-radix router designs are power-hungry with degraded performance, even though the hop count between PEs is reduced. Consequently, a topology-router co-design method for 3D NoC is desirable, so that the hop count between any two PEs and the radix of the 3D router design is as small as possible. Xu et al. [73] proposed a 3D NoC topology with low hop count (which is defined as low diameter) and low radix router design. The level 2D mesh is replaced with a network of long links connecting nodes that are at least m mesh-hops away, where m is a design parameter. In such a topology, long distance communications can leverage the long physical wire and vertical links to reach destination, achieving low total hop count, while the radix of the router is kept low. For application-specific NoC architecture, Yan et al. [79] also proposed a 3D-NoC synthesis algorithm that is based on a rip-up and reroute formulation for routing flows and a router merging procedure for network optimization to reduce the hop count.

6.3   3D OPTICAL NOC DESIGN

Aside from electrical NoC, 3D optical NoCs leveraging heterogeneous technology integration have also been proposed [86, 87]. The advancement in silicon photonics have motivated research on building on-chip optical interconnects with monolithic optical components, e.g., waveguides, light modulators and detectors, etc. However, building feasible optical on-chip networks requires massively fabricating these components with electrical devices. Since the optimal processes are different for optical and electrical devices, this presents an issue for technology tuning. This problem is resolved with 3D integration by using different technologies with different layers.

Ye et al. [86] have demonstrated an optical NoC using 3D-stacking, based on the Cygnus optical router architecture. Cygnus leverages waveguides and micro-ring resonators in the optical crossbar to switch data between directions. The micro-ring resonators can be viewed as couplers that can divert signals from waveguides. The micro-ring can be switched between the *on* and *off* states by electrical signals. In the *on* state it diverts optical signals while in the *off* state it lets optical signals pass by. Due to the absent of optical buffering, Cygnus proposes using circuit-switching and the optical path set-up and tear-down are done with an electrical hop-by-hop network. As discussed in the previous paragraph, however, it is technologically difficult to integrate the optical network with the electrical network. A solution surfaces with 3D integration, where the two networks are fabricated on different layers; control signals and inject/eject links are implemented by TSVs. Comparing 3D Cygnus with an electrical network having equal size, experiment results show up to 70% power saving and 50% latency saving. The drawback of the 3D Cygnus, however, is the slightly lower maximum throughput due to the overheads of circuit-switching.

Zhang et al. [87] yet propose another 3D optical NoC using packet switching. This work is motivated by Corona [88] which adapts token rings for optical interconnects. The problem with token rings, however, is the reduced utilization when the network size becomes large due to the elongated token circulating time. Zhang et al. propose to divide a large network into smaller regions and implement optical token rings inside and between regions. While this approach reduces token circulating time, it introduces a large number of waveguide crossings that severely affects optical signal integrity. To address this problem, they leverage the recently demonstrated "optical vias" that can couple light from one layer to another layer. They propose to stack crossing waveguides at different layers such that crosstalk and signal loss are largely reduced. As a result, the proposed 3D NoC architecture outperforms Corona and electrical networks in both performance and power. For example, the zero-load latency of the proposed 3D optical NoC is 28% lower than Corona, while the throughput is 2.4 times higher. In addition, the proposed 3D optical NoC consumes 6.5% less power than Corona.

6.4   IMPACT OF 3D TECHNOLOGY ON NOC DESIGNS

Since TSV vias contend with active device area, they impose constraints on the number of such vias per unit area. Consequently, the NoC design should be performed holistically in conjunction with other system components such as the power supply and clock network that will contend for the same interconnect resources.

The 3D integration using TSV (through-silicon-via) can be classified into one of the two following categories; (1) *monolithic approach* and the (2) *stacking approach*. The first approach involves a sequential device process, where the frontend processing (to build the device layer) is repeated on a single wafer to build multiple active device layers before the backend processing builds interconnects among devices. The second approach (which could be wafer-to-wafer, die-to-wafer, or die-to-die stacking) processes each active device layer separately using conventional fabrication techniques. These multiple device layers are then assembled to build up 3D ICs using bonding technology. Dies can be bonded face-to-face (F2F) or face-to-back (F2B). The microbump in face-to-face wafer bonding does not go through a thick buried Si layer and can be fabricated with a higher pitch density. In stacking bonding, the dimension of the TSVs. is not expected to scale at the same rate as feature size because alignment tolerance and thinned die/wafer height during bonding poses a limitation on the scaling of the vias.

The TSV (or micropad) size, length, and the pitch density, as well as the bonding method (face-to-face or face-to-back bonding, SOI-based 3D or bulk CMOS-based 3D) can have a significant impact on the 3D NoC topology design. For example, a relatively large size of TSVs. can hinder partitioning a design at very fine granularity across multiple device layers, and make the true 3D router design less possible. On the other hand, the monolithic 3D integration provides more flexibility in the vertical 3D connection because the vertical 3D via can potentially scale down with feature size due to the use of local wires for connection. Availability of such technologies makes it possible to partition the design at a very fine granularity. Furthermore, face-to-face bonding or SOI-based 3D integration may have a smaller via pitch size and higher via density than face-to-back bonding or bulk CMOS based integration. Such influence of the 3D technology parameters on the NoC topology design should be thoroughly studied and suitable NoC topologies for different 3D technologies should be identified with respect to the performance, power, thermal, and reliability optimizations.

CHAPTER 7

Thermal Analysis and Thermal-aware Design

Power and thermal issues have become the primary concerns in the traditional 2D IC design. Although emerging 3D technology offers several benefits over 2D, the stacking of multiple active layers in 3D design leads to higher power densities than its 2D counterpart, exacerbating the thermal issue. Therefore, it is essential to conduct thermal-aware 3D IC designs. This chapter presents an overview of thermal modeling for 3D IC and outlines solution schemes to overcome the thermal challenges.

One of the major concerns in the adoption of 3D technology is the increased power densities that can result from placing one power hungry block over another in the multi-layered 3D stack. Since the increasing power density and the resulting thermal impact are already major concerns in 2D ICs, the move to 3D ICs could accentuate the thermal problem due to increased power density, resulting in higher on-chip temperatures. High temperature has adverse impacts on circuit performance. The interconnect delay becomes slower while the driving strength of a transistor decreases with increasing temperature. Leakage power has an exponential dependence on the temperature and increasing on-chip temperature can even result in thermal runaways. In addition, at sufficiently high temperatures, many failure mechanisms, including electromigration (EM), stress migration (SM), time-dependent dielectric (gate oxide) breakdown (TDDB), and thermal cycling (TC), are significantly accelerated, which leads to an overall decrease in reliability. Consequently, it is very critical to model the thermal behaviors in 3D ICs and investigate possible solutions to mitigate thermal problems in order to fully take advantage of the benefits that 3D technologies offer.

7.1 THERMAL ANALYSIS

3D technology offers several benefits compared to traditional 2D technology. However, one of the major concerns in the adoption of 3D technology is the increased power densities that can result from placing one power hungry block over another in the multi-layered 3D stack. Since the increasing power density and the resulting thermal impact are already major concerns in 2D ICs, the move to 3D ICs could accentuate the thermal problem due to increased power density, resulting in higher on-chip temperatures. High temperature has adverse impacts on circuit performance, such as: (1) The interconnect delay becomes slower while the driving strength of a transistor decreases with increasing temperature; (2) leakage power has an exponential dependence on the temperature and increasing on-chip temperature can even result in thermal runaways. In addition, at sufficiently high temperatures, many failure mechanisms, including electromigration (EM), stress migration (SM), time-dependent dielectric (gate oxide) breakdown (TDDB), and thermal cycling (TC), are significantly accelerated, which leads to an overall decrease in reliability. Consequently, it is very critical to model the thermal behaviors in 3D ICs and investigate possible solutions to mitigate thermal problems in order to fully take advantage of the benefits that 3D technologies offer. This section surveys several thermal modeling approaches for 3D ICs. A detailed 3D modeling method is first introduced, and then compact thermal models are introduced.

3D Thermal Analysis based on Finite Difference Method or Finite Element Method.   Sapatnekar et al. proposed a detailed 3D thermal model [89]. The heat equation (7.1), which is a parabolic partial differential equation (PDE), defines on-chip thermal behavior at the macroscale:

$$\rho c_p \frac{\partial T(\mathbf{r}, t)}{\partial t} = k_t \nabla^2 T(\mathbf{r}, t) + g(\mathbf{r}, t), \tag{7.1}$$

where p represents the density of the material (in kg/m3), c_p is the heat capacity of the chip material (in J/(kg K)), T the temperature (in K), $\mathbf{r}$ is the spatial coordinate of the point where the temperature is determined, t is time (in sec), k_t is the thermal conductivity of the material (in W/(m K)), and g is the power density per unit volume (in W/m3). The solution of Eq. (7.1) is the transient thermal response. Since all derivatives with respect to time go to zeroes in the steady state, steady-state analysis is needed to solve the PDE, which is the well-known Poisson's equation.

A set of boundary conditions must be added in order to get a well-defined solution to Eq. (7.1). It typically involves building a package macro model and assuming a constant ambient temperature is interacted with the model. There are two methods to discretize the chip and to form a system of linear equations representing the temperature distribution and power density distribution: finite difference method (FDM) and finite element method (FEM). The difference between them is that FDM discretizes the differential operator while FEM discretizes the temperature field. Both of them can handle complicated material structures such as non-uniform interconnects distributions in a chip.

In FDM, a heat transfer theory is adopted, which builds an equivalent thermal circuit through the thermal-electrical analogy. The steady-state equation represents the network with thermal resistors connected between nodes and with thermal current sources mapping to power sources. The voltage and the temperature at the nodes can be computed by solving the circuit. The ground node is considered as a constant temperature node, typically the ambient temperature.

Since FDM methods are similar to power grid analysis problems, similar solution techniques can be used, such as the multigrid-based approaches. Li et al. [90] proposed multigrid (MG) techniques for fast chip level thermal steady-state and transient simulation. This approach avoids an explicit construction of the matrix problem that is intractable for most full-chip problems. Specific MG treatments are proposed to cope with the strong anisotropy of the full-chip thermal problem that is created by the vast difference inmaterial thermal properties and chip geometries. Importantly, this work demonstrates that only with careful thermal modeling assumptions and appropriate choices for grid hierarchy, MG operators, and smoothing steps across grid points, can a full-chip thermal problem be accurately and efficiently analyzed.

The FEM provides another avenue to solve Poisson's equation. In finite element analysis, the design space is first discretized or meshed into elements. Different element shapes can be used such as tetrahedra and hexahedra. For the on-chip problem, where all heat sources are modeled as being rectangular, a reasonable discretization for the FEM divides the chip into 8-node rectangular hexahedral elements [91]. The temperatures at the nodes of the elements constitute the unknowns that are computed during finite element analysis, and the temperature within an element is calculated using an interpolation function that approximates the solution to the heat equation within the elements.

Compact 3D IC Thermal Modeling.   3D IC thermal analysis with FDM or FEM can be very time consuming and therefore is not suitable to be used for design space exploration where the thermal analysis has to be performed iteratively. Therefore a compact thermal model for 3D IC is desirable. For traditional 2D design, one widely used compact thermal model is called HotSpot [92], which is based on an equivalent circuit of thermal resistances and capacitances that correspond to microarchitecture components and thermal packages. In a well-known duality between heat transfer and electrical phenomena, heat flow can be described as a "current" while temperature difference is analogous to a "voltage." The temperature difference is caused by heat flow passing through a thermal resistance. It is also necessary to include thermal capacitances for modeling transient behavior to capture the delay before the temperature reaching steady state due to a change in power consumption. Like electrical RC constants, the thermal RC time constants characterize the rise and fall times led by the thermal resistances and capacitances. In the rationale, the current and heat flow are described by exactly the same different equations for a potential difference. These equivalent circuits are called compact models or dynamic compact models if thermal capacitors are included. For a microarchitecture unit, the dominant mechanism to determine the temperature is the heat conduction to the thermal package and to neighboring units.

In HotSpot, the temperature is tracked at the granularity of individual microarchitectural units and the equivalent RC circuits have at least one node for each unit. The thermal model component values do not depend on initial temperature or the particular configurations. HotSpot is a simple library that generates the equivalent RC circuit automatically and computes temperature at the center of each block with power dissipation over any chosen time step.

Based on the original HotSpot model, a 3D thermal estimation tool named HS3D was introduced [93]. HS3D allows 3D thermal evaluation despite a largely unchanged computational model and methods in HotSpot. The inter-layer thermal vias can be approximated by changing the vertical thermal resistance of the materials. HS3D library allows incompletely specified floorplans as input and ensures accurate thermal modeling of large floorplan blocks. Many routines have been recreated for optimizations of loop accesses, cache locality, and memory paging. These improvements offer reduced memory usage and runtime reduction by over three orders when simulating a large number of floorplans. To guarantee the correctness and efficiency of HS3D library, the comparison between this new library and a commercial FEM software was performed. First a 2D sample device and package is used for the verification. The difference of the average chip temperatures from HS3D and FEM software is only 0.02 °C. Multi-layer (3D) device modeling verification is provided using 10 μm thick silicon layers and 2 μm thick interlayer material. The test case includes two layers with a sample processor in each layer. The experiment results show that the average temperature mis-estimation is 3 °C. In addition, the thermal analysis using FEM software costs seven minutes while only costs one second HS3D. It indicates that HS3D provides not only high accuracy but also high performance (low run time). The extension in HS3D was integrated into HotSpot in the later versions to support 3D thermal modeling.

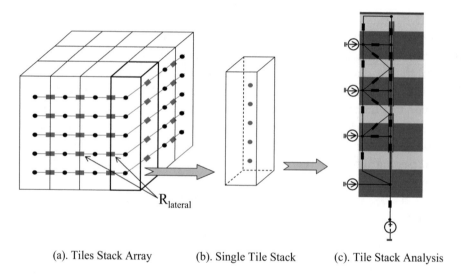

(a). Tiles Stack Array   (b). Single Tile Stack   (c). Tile Stack Analysis

Figure 7.1: Resistive thermal model for a 3-D IC [94].

Cong and Zhang derived a closed-form compact thermal model for thermal via planning in 3D ICs [94]. In their thermal resistive model, a tile structure is imposed on the circuit stack with each tile the size of a via pitch, as shown in Fig. 7.1(a). Each tile stack contains an array of tiles, with one tile for each device layer, as shown in Fig. 7.1(b). A tile either contains one via at the center, or no via at all. A tile stack is modeled as a resistive network, as shown in Fig. 7.1(c). A voltage source is utilized for the isothermal base, and current sources are present in each silicon layer to represent heat sources. The tile stacks are connected by lateral resistances. The values of the resistances in the network are determined by a commercial FEM-based thermal simulation tool.

To further improve analysis accuracy without losing efficiency, Yang et al. proposed an incremental and adaptive chip-package thermal analysis flow named ISAC [95]. During thermal analysis, both time complexity and memory usage are linearly or superlinearly related to the number of thermal elements. ISAC incorporates an efficient technique for adapting thermal element spatial resolution during thermal analysis. This technique uses incremental refinement to generate a tree of heterogeneous rectangular parallelepipeds that supports fast thermal analysis without loss of accuracy. Within ISAC, this technique is incorporated with an efficient multigrid numerical analysis method, yielding a comprehensive steady-state thermal analysis solution.

7.2   THERMAL-AWARE FLOORPLANNING FOR 3D PROCESSORS

Physical design tools play an important role in 3D design automation since one essential difference between 2D and 3D CAD tools is the physical design. The physical design tools for 3D IC can be classified into three categories: floorplanning, placement, and routing. In addition, one major concern in 3D ICs is the increased power density due to placing one thermal critical block over another in the 3D stack. Consequently, thermal-aware physical design tools should be developed to address both thermal and physical tool issues.

The physical design methodologies for 3D IC depend on the partitioning strategy, which is determined by the 3D technology process. As discussed in Chapter 1, the partitioning strategy in 3D ICs can be classified into two categories: *fine-granularity partitioning* and *coarse-granularity partitioning*. For monolithic 3D approach (such as the MLBS approach) or TSV-based 3D ICs with extremely small TSVs, it is possible to perform fine-granularity partitioning with the gate-level 3D placement and routing; on the other hand, with large TSV size (with via diameter larger than 1 μm), it makes more sense to perform coarse-granularity partitioning at the unit level or core level, to avoid large area overhead. For coarse-granularity partitioning, since each unit or core is still a 2D design, the physical placement and routing for each unit/core can still be done with conventional placement and routing EDA tools, and then for coarse-granularity 3D partitioning, only 3D floorplanning is needed. Consequently, 3D floorplanning is the most needed physical design tool for near term 3D IC design. In this section, we will first describe in detail a thermal-aware 3D floorplanning framework [96], and then give a quick survey on thermal-aware 3D placement and routing.

3D IC design is fundamentally related to the topological arrangement of logic blocks. Therefore, physical design tools play an important role in the adoption of 3D technologies. New placement and routing tools are necessary to optimize 3D instantiations and new design tools are required to optimize interlayer connections. A major concern in the adoption of 3D architecture is the increased power densities that can result from placing one computational block over another in the multilayered 3D stack. Since power densities are already a major bottleneck in 2D architectures, the move to 3D architectures could accentuate the thermal problem. Even though 3D chips could offer some respite due to reduced interconnect power consumption (as a result of the shortening of many long wires), it is imperative to develop thermally aware physical design tools—for example, partition design to place highly loaded, active gates in layers close to the heat-sink.

Tools for modeling thermal effects on chip-level placement have been developed. Recently, Cong et al. [97] proposed a thermal-driven floorplanning algorithm for 3D ICs. Chu and Wong [98] proposed using a matrix synthesis problem (MSP) to model the thermal placement problem. A standard cell placement tool to even thermal distribution has been introduced by Tsai and Kang [99] with their proposed compact finite difference method (FDM) based temperature modeling. In [91], thermal effect was formulated as another force in a force-directed approach to direct the placement procedure for a thermally even standard cell placement. Another design metric, reliability, was taken care of in [100] when doing a multi-layer System-on-Package floorplanning; thermal issue was neglected, though.

This section presents a thermal-aware floorplanner for a 3D microprocessors [96]. The floorplanner is unique in that it accounts for the effects of the interconnect power consumption in estimating the peak temperatures. We describe how to estimate temperatures for 3D microprocessors and show the effectiveness of the thermal-aware tool in reducing peak temperatures using one microprocessor design and four MCNC benchmarks.

B*-tree Floorplan Model

The floorplanner used in this work is based on the B*-tree representation, which was proposed by Chang et al. [101]. A B*-tree is an ordered binary tree which can represent a non-slicing admissible floorplan. Figure 7.2 shows a B*-tree representation and its corresponding floorplan. The root of a B*-tree is located at the bottom-left corner. A B*-tree of an admissible floorplan can be obtained by depth-first-search procedure in a recursive fashion. Starting from the root, the left subtree is visited first then followed by the right subtree, recursively. The left child of node n_i is the lowest unvisited module in R_i, which denotes the set of modules adjacent to the right boundary of module b_i. On the other hand, the module, representing the right child of n_i, is adjacent to and above module b_i. We assume the coordinates of the root node are always (0,0) residing at bottom-left corner. The geometric relationship between two modules in a B*-tree is maintained as follows. The x-coordinate of node n_j is $x_i + w_i$, if node n_j is the left child of node n_i. That is, b_j is located and is adjacent to the right-hand side of b_i. For the right child n_k of n_i, its x-coordinate is the same as that of n_i, with module b_k sitting adjacent to and above b_i. While the original B*-tree structure was developed and used for the 2D floorplanning problem, we extend this model to represent a multi-layer 3D floorplanning (see Fig. 7.3) and modify the perturbation function to handle 3D floorplans in this work.

There are six perturbation operations used in our algorithm and they are listed below:

(1) Node swap, which swaps two modules.

(2) Rotation, which rotates a module.

(3) Move, which moves a module.

(4) Resize, which adjusts the aspect-ratio of a soft module.

(5) Interlayer swap, which swaps two modules at different layers.

(6) Interlayer move, which moves a module to a different layer.

The first three perturbations are the original moves defined in [101]. Since these moves only have influence on the floorplan in single layer, more interlayer moves, (5) and (6) are needed to explore the 3D floorplan solution space.

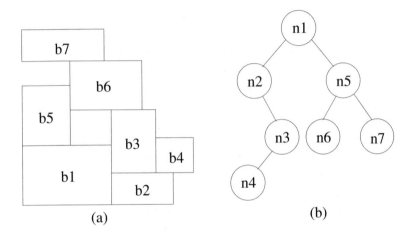

Figure 7.2: (a) An example floorplan; (b) the corresponding B*-tree [101].

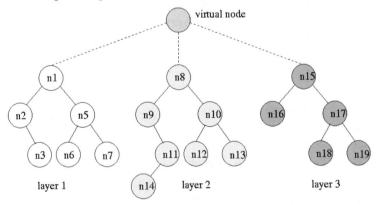

Figure 7.3: Extending B*-tree model to represent 3D floorplanning [5].

Simulated Annealing Engine

A simulated annealing engine is used to generate floorplanning solutions. The inputs to our floor-planning algorithm are the area of all functional modules and interconnects among modules. However, the actual dimension of each module is unknown *a priori* except for its area before placement. That is, we have to treat them as *soft* modules. Thus, we provide the choice of adjusting aspect-ratio as one perturbation operation. During simulated process, each module dynamically adjusts its aspect-ratio to fit closely with the adjacent modules, that is, with no dead space between two modules. A traditional weighted cost representing optimization costs (area and wire length) is generated after each perturbation.

Different from 2D floorplanning, our 3D floorplanner uses a two-stage approach. The first stage tries to partition the blocks to appropriate layers, and tries to minimize the packed area difference between layers and total wire length using all perturbation operations (one through six listed in previous subsection). However, because the first stage tries to balance the packed areas of the different layers, the floorplan of some layers may not be compactly packed. The second stage is intended to overcome this problem. Thus, in the second stage, we start with the partitioning solution generated by the first stage and focus on adjusting the floorplan of each layer simultaneously with the first four operations. At this point, there are no inter-layer operations to disturb module partition of each layer obtained from stage one.

One problem of 3D floorplanning is the final packed area of each layer must match to avoid penalties of chip area. For example, if the final width of packed modules of layer $L1$ is larger than the final width of packed modules in layer $L2$ and the height of L1 is smaller that of L2, a significant portion of chip area is wasted due to the need for the layer dimensions to match for manufacturing. To make sure the dimension of each layer will be *compatible*, we adopt the concept of dimension deviation $dev(F)$ in [100]. The goal is to minimize $dev(F)$, which tells the deviation of the upper-right corner of a floorplan from the average Ave_x, Ave_y values. The value, Ave_x can be calculated by $\sum ux(f_i)/k$, where $ux(f_i)$ is the x-coordinate of upper-right corner of floorplan i, and k indicates the number of layers. The value Ave_y can be obtained in a similar manner. Thus, $dev(F)$ is formulated as $\sum_i^{\# \ layers} |Ave_x - ux(f_i)| + |Ave_y - uy(f_i)|$. The modified cost function for 3D floorplanner can be written as

$$cost = \alpha * area + \beta * wl + \gamma * dev(F) \tag{7.2}$$

where *area*, *wl* are chip area and wire length, respectively.

Temperature Approximation

Although HS3D [93] can be used to provide temperature feedbacks, when evaluating a large number of solutions during simulated procedure, it is not wise to involve the time-consuming temperature calculation every time. Other than using the actual temperature values, we have adopted the power density metric as a thermal-conscious mechanism in our floorplanner. The temperature is heavily dependent on power density based on a general temperature-power equation: $T = P * R = P * (t/k * A) = (P/A) * (t/k) = d * (t/k)$, where t is the thickness of the

Table 7.1: Floorplanning results of 2D architecture

Circuit	2D			2D(thermal)		
	wire (um)	area (mm²)	peakT	wire (um)	area (mm²)	peakT
Alpha	339672	29.43	114.50	381302	29.68	106.64
xerox	542926	19.69	123.75	543855	19.84	110.45
hp	133202	8.95	119.34	192512	8.98	116.91
ami33	44441	1.21	128.21	51735	1.22	116.97
ami49	846817	37.43	119.42	974286	37.66	108.86

chip, k is the thermal conductivity of the material, R is the thermal resistance, and d is the power density. Thus, we can substitute the temperature and adopt the power density, according to the equation above, to approximate the 3-tie temperature function, $C_T = (T - T_o)/T_o$, proposed in [97] to reflect the thermal effect on a chip. As such, the 3-tie power density function is defined as $P = (P_{max} - P_{avg}))/P_{avg}$, where P_{max} is the maximum power density while P_{avg} is the average power density of all modules. The cost function for 2D architecture used in simulated annealing can be written as

$$cost = \alpha * area + \beta * wl + \gamma * P. \qquad (7.3)$$

For 3D architectures, we also adopt the same temperature approximation for each layer as horizontal thermal consideration. However, since there are multiple layers in 3D architecture, the horizontal consideration alone is not enough to capture the coupling effect of heat. The vertical relation among modules also needs to be involved and is defined as: $OP(TPm) = \sum(Pm + Pm_i) * overlap\_area$, where $OP(TPm)$ stands for the summation of the power density of module, Pm, and all overlapping module m_i with module m and their relative power densities multiplying their corresponding overlapped area.

The rationale behind this is that for a module with relatively high power density in one layer, we want to minimize its accumulated power density from overlapping modules located in different layers. We can define the set of modules to be inspected, so the total overlap power density is $TOP = \sum OP(TPi)$, for all modules in this set. The cost function for 3D architecture is thus modified as follows:

$$cost = \alpha * area + \beta * wl + \phi * dev(F) + \gamma * P + \delta * TOP. \qquad (7.4)$$

At the end of algorithm execution, the actual temperature profile is reported by our HS3D tool.

Experiment Results

We implemented the proposed floorplanning algorithm in C++. The thermal model is based on the HS3D. In order to effectively explore the architecture-level interconnect

Table 7.2: Floorplanning results of 3D architecture

Circuit	3D			3D(thermal)		
	wire (um)	area (mm²)	peakT	wire (um)	area (mm²)	peakT
Alpha	210749	15.49	135.11	240820	15.94	125.47
xerox	297440	9.76	137.51	294203	9.87	127.31
hp	124819	4.45	137.90	110489	4.50	134.39
ami33	27911	0.613	165.61	27410	0.645	155.57
ami49	547491	18.55	137.71	56209	18.71	132.69

power consumption of a modern microprocessor, we need a detailed model which can act for the current-generation high-performance microprocessor designs. We have used IVM (http://www.crhc.uiuc.edu/ACS/tools/ivm), a Verilog implementation of an Alpha-like architecture (denoted as Alpha in the rest of this section) at register-transfer-level, to evaluate the impacts of both interconnect and module power consumptions at the granularity of functional module level. A diagram of the processor was shown in Fig. 7.4. Each functional block in Fig. 7.4 represents a module used in our floorplanner. The registers between pipeline stages are also modeled but not shown in the figure.

The implementation of microprocessor has been mapped to a commercial 160 nm standard cell library by Design Compiler and placed and routed by First Encounter under 1GHz performance requirement. There are in total 34 functional modules and 168 netlists extracted from the processor design. The area and power consumptions from the actual layout served as inputs to our algorithm. Other than Alpha processor, we have also used MCNC benchmarks to verify our approach. A similar approach in [99] is used to assign the average power density for each module in the range of $2.2*10^4$ (W/m²) and $2.4*10^6$ (W/m²). The total net power is assumed to be 30% of total power of modules due to lack of information for the MCNC benchmarks, and the total wire length used to be scaled during floorplanning is the average number from 100 test runs with the consideration of area factor alone. The widely used method of half-perimeter bounding box model is adopted to estimate the wire length. Throughout the experiments, two-layer 3D architecture was assumed due to a limited number of functional modules and excessively high power density beyond two layers; however, our approach is capable of dealing with multiple-layer architecture.

Tables 7.1 and 7.2 show the experiment results of our approach when considering traditional metrics (area and wire) and thermal effect. When taking thermal effect into account, our thermal-aware floorplanner can reduce the peak temperature by 7% on average while increasing wirelength by 18% and providing a comparable chip area as compared to the floorplan generated using traditional metrics.

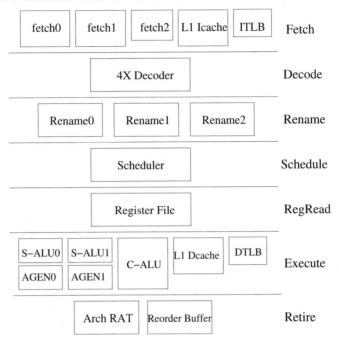

Figure 7.4: Processor model diagram [96].

When we move to 3D architectures, the peak temperatures increased by 18% (on average) as compared to the 2D floorplan due to the increased power density. However, the wire length and chip area reduced by 32% and 50%, respectively. The adverse effect of the increased power density in the 3D design can be mitigated by our thermal-aware 3D floorplanner, which lowers the peak temperature by an average of 8 degrees with little area increase as compared to the 3D floorplanner that does not account for thermal behavior. As expected, the chip temperature is higher when we step from 2D to 3D architecture without thermal consideration. Although the wire length is reduced when moving to 3D and thus accordingly reduces interconnect power consumption, the temperature for 3D architecture is still relatively high due to the accumulated power densities and smaller chip footprints. After applying our thermal-aware floorplanner, the peak temperature is lowered to a moderate level through the separation of high power density modules in different layers.

7.3   THERMAL-HERDING: THERMAL-AWARE ARCHITECTURE DESIGN

In this section, a family of thermal herding, microarchitecture techniques for controlling Hotspots in high performance 3D processors, is presented. Thermal herding techniques herd or steer the switching activity in the 3D processor to the die that is closest to the heat sink and reduce the total power and power density while still maintaining the performance benefits. It adopts different strategies for different components in the processor such as placing the frequently switching 16-bits on the top die in the significance-partitioned datapath, proposing a 3D-aware allocation scheme for instruction scheduler, an address memorization approach for the load and store queues, a partial value encoding for the L1 data cache, and exploiting a form of frequent partial value locality in target addresses for a branch target buffer.

Thermal herding is proposed based on some observations made by prior work that only a few of the least significant bits are needed in the data used in many integer instructions [102], i.e., many 64-bit integer values only requires 16 or fewer bits for the representation. In addition, it is highly predictable for an instruction's usage of low-width values. Therefore, the datapath can be organized by assigning each 16 bits to a separate dies (four dies) in 3D stacking. The least significant bits can be placed on the top layer which is closest to the heat sink so that the power on the other three dies may be saved using the instruction's width prediction. A prediction is made for each instruction whether to use low-width ($\leq$16bits) or full-width ($\geq$ 16bits) values. A simple program counter indexed two-bit saturating counter predictor is used. When the predictor indicates the data is low-width usage but it is actually a full-width usage then the result is an unsafe misprediction, which requires pipeline stalls in related stages. Some critical microarchitecture components are discussed as follows.

Register Files: Each 64-bit entry in the register file is partitioned into four layers with the least significant bits closest to the heat sink. The width prediction is used to determine gating control signals for other three layers before the actual register file access. If low-width instruction is predicted then only the top layer portion of the register file is active. In this case, the power density is similar to that of a 2D register file design. A width memorization bit is also provided for each entry in the top layer, indicating whether the remaining three layers contain non-zero values. The processor compares this bit to the predicted width when reading it. If the width prediction is low and the actual width is full, then the processor stalls the previous stages of the register file and activates the logic on the other three layers. It also corrects the prediction to prevent further stalls.

Arithmetic Units: Only 3D integer adder is presented but the concept can be extended to other arithmetic units. A tree-based adder is partitioned into four layers with the least significant bits residing on the top layer. The processor uses predication information to decide if the clock gate should be activated for other three layers. Two possible unsafe width mispredictions should be handled properly. One is misprediction on an instruction's input operands. In this scenario, since the arithmetic unit is not fully active at the start of execution, one cycle stall is needed to activate the upper 48 bits. Another scenario is misprediction on the output, in which the width misprediction is not known at the beginning of the computation. The instruction needs to be executed again to guarantee the correctness, causing performance penalty. Therefore, the accuracy of the width predictor is very important.

Bypass Network: The bypass network does not need additional circuitry to handle the width prediction since the unsafe misprediction is handled by the arithmetic units. With a correct low-width prediction, only the drivers/wires on the top layer consume dynamic power. In addition, the partitioning in 3D reduces the wire length so that the latency and power are reduced accordingly.

Instruction Scheduler: In the instruction scheduler, there is one reservation station (RS) for each instruction dispatched but not executed. When an instruction is ready to issue, the instruction broadcasts its destination identifier to notify dependent instructions. The instruction scheduler is partitioned based on the RS entries so that the length of the broadcast buses is significantly reduced, resulting in latency and power reduction. A modified allocation algorithm is also adopted to move instructions to the top layer so that the active entries are closer to the heat sink. If there are no available entries in the top layer, the allocator will check the layer that is next closest to the heat sink.

Load and Store Queues: The load and store queues are used to track the data and addresses of instructions. The queues are partitioned in a similar fashion with the main datapath due to its similarity with register file. Load and store addresses are normally full-width values but the upper bits of the addresses do not change frequently. To take advantage of this feature, a partial address memorization (PAM) scheme is proposed. The low-order 16 bits of a load or store's address is broadcast on the top layer. In addition, whether the remaining 48 bits are the same as the most recent store address is also indicated. To summarize, the PAM approach tries to herd the address broadcasts and comparisons to the top layer.

Data Cache: The L1 data cache is organized in a word-partitioned manner due to its similarity with register file. Memorization bits are also provided to detect unsafe width mispredictions. In addition, the definition of a "low-width" value for load and store instructions is broadened to increase the frequency of low-width values. Two bits instead of one single width memorization bit are stored to encode the upper 48 bits. Value "00" means the upper 48 bits are all zeros. Value "01" indicates all ones. Value "10" means the upper bits are identical to those of the referencing address. Value "11" means the upper bits are not encodeable.

Front End: Data-centric approaches for partitioning the front end are not effective since no data values are handled. A register alias table (RAT) is implemented to place the ports of each instruction on different layers so that unnecessary die-to-tie vias are avoided. The instruction that requires the most register name comparisons is placed on the top layer so that most switching activities are moved to the top layer. For branch predictor based on two-bit saturating counters, the counters are partitioned into two separate arrays: direction bit array and hysteresis array. The more frequently used direction-bit array is place on the layer closer to the heat sink. For the branch target buffers (BTBs), since most branch targets are located relatively close to the originating branch, they are organized like the data cache. The low-order 16 bits are placed on the top layer with one extra target memorization bit indicating whether the bits on the other three layers should be accessed.

3D thermal herding techniques can improve IPC by reducing the pipeline depth and L2 latency but it also can reduce IPC due to width mispredictions. By conducting the experiments on several benchmarks, the overall performance is improved since the pipeline reduction benefits outweigh the performance penalties caused by width mispredictions. The power consumption is also reduced due to two reasons: the length of wires is reduced because of 3D stacking and thermal herding reduces the switching activities because of clock gating. The thermal experiments show that thermal herding techniques successfully control power density and mitigate 3D thermal issues.

Conclusion.   Increasing power density in 3D ICs can result in higher on-chip temperatures, which can have a negative impact on performance, power, reliability, and the cost of the chip. Consequently, thermal modeling and thermal-aware design techniques are very critical for future 3D architecture designs. This chapter presents an overview of thermal modeling for 3D IC and outlines architecture design schemes to overcome the thermal challenges.

CHAPTER 8

Cost Analysis for 3D ICs

The majority of the 3D processor design research has focused on how to take advantage of the performance, power, smaller form-factor, and heterogeneous integration benefits offered by 3D integration. However, when it comes to the discussion on the adoption of such emerging technology as a mainstream design approach, it all comes down to the question of the 3D integration cost. *All the advantages of 3D ICs ultimately have to be translated into cost savings when a design strategy has to be decided.* For example, designers may ask themselves questions like,

- *Do all the benefits of 3D IC design come with a much higher cost?* For example, 3D bonding incurs extra process cost, and the Through-Silicon Vias (TSVs) may increase the total die area, which has a negative impact on the cost; however, smaller die sizes in 3D ICs may result in higher yield than that of a larger 2D die, and reduce the cost.

- *How to do 3D integration in a cost-effective way?* For example, to re-design a small chip may not gain the cost benefits of improved yield resulted from 3D integration. In addition, if a chip is to be implemented in 3D, how many layers of 3D integration would be cost effective? and should one use wafer-to-wafer or die-to-wafer stacking [5]?

- *Are there any design options to compensate the extra 3D bonding cost?* For example, in a 3D IC, since some global interconnects are now implemented by TSVs, it may be feasible to use fewer number of metal layers for each 2D die. In addition, heterogeneous integration via 3D could also help cost reduction.

Cost analysis for 3D ICs at the early design stage is critical to answer these questions. Considering 3D integration has both positive and negative effects on the manufacturing cost, there are no clear answers yet whether 3D integration is cost-effective or how to achieve 3D integration in a cost-effective way. In order to maximize the positive effect and compensate the negative effect on cost, it becomes critical to analyze the 3D IC cost at the early design stage. This early stage cost analysis can help the chip designers make a decision on whether 3D integration should be used, or which 3D partitioning strategy should be adopted. *Cost-efficient design is the key for the future wide adoption of the emerging 3D IC design, and 3D IC cost analysis needs close coupling between the 3D IC design and 3D IC process.*[1]

---

[1]IC cost analysis needs a close interaction between designers and foundry. We work closely with our industrial partners to perform 3D IC cost analysis. However, we cannot disclose absolute numbers for the cost, and therefore in this book, we either use arbitrary units (a.u.) or normalized value to present the data.

In this chapter, we first propose a 3D IC cost analysis methodology with a complete set of cost models that include wafer cost, 3D bonding cost, package cost, and cooling cost. Using this cost analysis methodology along with the existing performance, power, area estimation tools, such as McPAT [103], we estimate the 3D IC cost in two cases: one is for fully customized ASICs, the other is for many-core microprocessor designs. During these case studies, we explore some guidelines showing what a cost-effective 3D IC fabrication option should be in the future.

Note that 3D integration is not yet a mature technology with very well-developed and tested cost models; the optimal condition concluded by this work is subject to parameter changes. However, the major contribution of this work is to provide a cost estimation methodology for 3D ICs. To our best knowledge, this is the first effort to model the 3D IC cost with package and cooling cost included, while the majority of the existing 3D IC research activities mainly focused on the circuit performance and power consumption.

In this chapter, we propose a complete set of 3D cost models, which not only includes the wafer cost and the 3D bonding cost, but also includes another two critical cost components, the package cost and the cooling cost. Both of them should not be ignored because while 3D integration can potentially reduce the package cost by having a smaller chip footprint, the multiple 3D-stacked dies might increase the cooling cost due to the increased power density. By using the proposed cost model set, we conduct two case studies on fully customized ASIC and many-core microprocessor designs, respectively. The experimental results show that our 3D IC cost model makes it feasible to estimate the final system-level cost of the target chip design at the early design stage. More importantly, it gives some guidelines to determine what a cost-effective 3D fabrication option should be.

8.1   3D COST MODEL

It is estimated that design choices made within the first 20% of the total design cycle time will ultimately result in up to 80% of the final product cost [104]. Hence, to facilitate the design decision of using 3D integration from a cost perspective, it is necessary to perform cost analysis at the early design stage. While it is straightforward to sum up all the incurred costs after production, predicting the final chip cost at the early design stage is a big challenge because most of the detailed design information is not available at that stage. Taking two different IC design styles (*ASIC design style and many-core microprocessor design style*) as examples:

- *ASIC Design style:* At the early design stage, probably only the system-level (such as behavioral level or RTL level) design specification is available, and a rough gate count can be estimated from a quick high-level synthesis or logic synthesis tools, or simply by the past design experience.

- *Microprocessor design style:* Using a many-core microprocessor design as an example, the information available at the early design stage is also very limited. The design specification may just include information as follows: (1) The number of microprocessor cores; (2) the type of microprocessor cores (e.g., in-order or out-of-order); and (3) the number of cache levels and the cache capacity of each level. All these specifications are on the architectural level. Referring to previous design experience, it is feasible to achieve a rough estimation on the gate count for logic-intensive cores and the cell count for memory-intensive caches, respectively.

Consequently, it is very likely that at the early design stage, the cost estimation is simply based on the design style and a rough gate count for the design as the initial starting point. In this section, we describe how to translate the logic gate count (or the memory cell count) into higher level estimations, such as the die area, the metal layer count, and the TSV count, given a specified fabrication process node. The die area estimator and the metal layer estimator are two key components in our cost model: (i) larger die area usually causes lower die yield, and thus leads to higher chip cost; (ii) the fewer metal layers are required, the less the number of fabrication steps (and fabrication masks) are needed, which reduce the chip cost. It is also important to note that the 3D partitioning strategy can directly affect these two numbers: (1) 3D integration can partition the original 2D design into several smaller dies; (2) TSVs can potentially shorten the total interconnect length and thus reduce the number of metal layers.

Die Area Estimation.   At the early design stage, the relationship between the die area and the gate count can be roughly described as follows,

$$A_{\text{die}} = N_{\text{gate}} A_{\text{gate}} \tag{8.1}$$

where N_{gate} is the gate count and A_{gate} is an empirical parameter that shows the proportional relationship between area and gate counts. Based on empirical data from many industrial designs, in this work, we assume that A_{gate} is equal to $3125\lambda^2$, in which λ is half of the feature size for a specific technology node. Although this area estimation methodology is straightforward and highly simplified, it accords with the real-world measurement quite well.[2]

Wire Length Estimation.   As the only inputs at the early design stage are the estimation of gate counts, it is necessary to further have an estimation on the design complexity, which can be represented by the total length of wire interconnects. *Rent's Rule* [105] is a well-known and powerful tool that reveals the trend between the number of signal terminals and the number of internal gates. *Rent's Rule* can be expressed as follows,

$$T = k N_{\text{gate}}^p \tag{8.2}$$

where the parameters k and p are Rent's coefficient and exponent and T is the number of signal terminals. Although *Rent's Rule* is an empirical result based on the observations of previous designs and it is not proper to use it for non-traditional designs, it does provide a useful framework to compare similar architecture and it is plausible to use it as a part of our 3D IC cost model.

---

[2]Note that it is necessary to update A_{gate} along with the technology advance because the gate length is not strictly linearly scaled.

Based on Rent's Rule, Donath discovered it can be used to estimate the average wire length [106] and Davis et al. found it can be used to estimate the wire length distribution in VLSI chips [107]. As a result, in this work, we use the derivation of *Rent's Rule* to predict the wire length distribution function $i(l)$ [107], which has the following forms,

Region I: $1 \leq l \leq \sqrt{N_{\text{gate}}}$

$$i(l) = \frac{\alpha k}{2} \Gamma \left(\frac{l^3}{3} - 2\sqrt{N_{\text{gate}}} l^2 + 2N_{\text{gate}} l \right) l^{2p-4}$$

Region II: $\sqrt{N_{\text{gate}}} \leq l < 2\sqrt{N_{\text{gate}}}$

$$i(l) = \frac{\alpha k}{6} \Gamma \left(2\sqrt{N_{\text{gate}}} - l \right)^3 l^{2p-4} \tag{8.3}$$

where l is the interconnect length in units of the gate pitch, α is the fraction of the on-chip terminals that are sink terminals and is related to the average fanout of a gate ($f.o.$) as follows,

$$\alpha = \frac{f.o.}{f.o. + 1} \tag{8.4}$$

and Γ is calculated as follows,

$$\Gamma = \frac{2N_{\text{gate}} \left(1 - N_{\text{gate}}^{p-1} \right)}{\left(-N_{\text{gate}}^p \frac{1+2p-2^{2p-1}}{p(2p-1)(p-1)(2p-3)} - \frac{1}{6p} + \frac{2\sqrt{N_{\text{gate}}}}{2p-1} - \frac{N_{\text{gate}}}{p-1} \right)}. \tag{8.5}$$

Metal Layer Estimation.   After estimating the die area and the length of wire, we are able to further predict the number of metal layers that are required to route all the interconnects within the die area constraint. The number of required metal layers for routing depends on the complexity of the interconnects. A simplified metal layer estimation can be derived from the average wire length [104] as follows,

$$n_{\text{wire}} = \frac{f.o.\bar{R}_m w}{\eta} \sqrt{\frac{N_{\text{gate}}}{A_{\text{die}}}} \tag{8.6}$$

where $f.o.$ refers to the average gate fanout, w to the wire pitch, η to the utilization efficiency of metal layers, $\bar{R}_m$ to the average wire length, and n_{wire} to the number of required metal layers.

Such a simplified model is based on the assumptions that each metal layer has the same utilization efficiency and the same wire width [104]. However, such assumptions may not be valid in real design [108]. To improve the estimation of the number of metal layers needed for feasible routing, a more sophisticated metal layer estimation method is derived from the wire length distribution rather than the simplified average wire length estimation. The basic flow of this method is explained as follows,

- *Estimate the available routing area of each metal layer with the expression:*

$$K_i = \frac{A_{\text{die}}\eta_i - 2Av_i \left(N_{\text{gate}} \, f.o. - I(l_i) \right)}{w_i} \qquad (8.7)$$

where i is the metal layer, η_i is the layer's utilization efficiency, w_i is the layer's wire pitch, Av_i is the layer's via blockage area, and function $I(l)$ is the cumulative integral of the wire length distribution function $i(l)$, which is expressed in Eq. 8.3.

- *Assume that shorter interconnects are routed on lower metal layers. Starting from Metal 1, we route as many interconnects as possible on the current metal layer until the available routing area is used up.* The interconnects routed on each metal layer can be express as:

$$\chi L(l_i) - \chi L(l_{i-1}) \leq K_i \qquad (8.8)$$

where $\chi = 4/(f.o. + 3)$ is a factor accounting for the sharing of wires between interconnects on the same net [107, 109]. The function $L(l)$ is the first-order moment of $i(l)$.

- *Repeat the same calculations for each metal layer in a bottom-up manner until all the interconnects are routed properly.*

By applying the estimation methodology introduced above, we can predict the die area and the number of metal layers at the early design stage where we only have the gate count as the input. Table 8.1 lists the values of all the related parameters [108, 110]. Figure 8.1 shows an example which estimates the area and the number of metal layers of 65nm designs with different scale of gates.

Figure 8.1 also shows an important implication for 3D IC cost reduction: *When a large 2D chip is partitioned into multiple smaller dies with 3D stacking, each smaller die requires fewer number of metal layers to satisfy the interconnect routability requirements.* Such metal layer reduction opportunity can potentially offset the cost of extra steps caused by 3D integration such as 3D bonding and test.

TSV Estimation.   The existence of TSVs in the 3D IC have its effects on the wafer cost as well. These effects are twofold,

1. In 3D ICs, some global interconnects are now implemented by TSVs, going between stacked dies. This could lead to the reduction of the total wire length, and provides opportunities for metal layer reduction for each smaller die;

2. On the other hand, 3D stacking with TSVs may increase the total die area, since the silicon area where TSVs punch through may not be utilized for building devices or 2D metal layer connections.[3]

---

[3]Based on current TSV fabrication technologies, the diameter of TSVs ranges from $0.2\mu m$ to $10\mu m$ [113]. In this work, we use the TSV diameter of $10\mu m$, and assume that the keepout area diameter is 2.5X of TSV diameter, which is $25\mu m$.

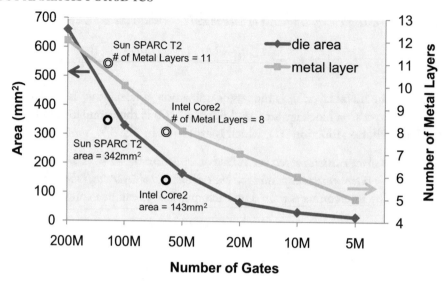

Figure 8.1: An example of early design estimation of the die area and the metal layer under 65nm process [111]. This estimation is well correlated with the state-of-the-art microprocessor designs. For example, Sun SPARC T2 [112] contains about 500 million transistors (roughly equivalent to 125 million gates), with an area of 342mm² and 11 metal layers).

While the first effect is already explained by Fig. 8.1 showing the existence of TSVs lead to a potential metal layer reduction on the horizontal direction, modeling the second effort needs extra efforts by adding the estimation of the number of TSVs and their associated silicon area overhead.

To predict the number of required TSVs for a certain partition pattern, a derivation of *Rent's Rule* [106] describing the relationship between the interconnect count (X) and the gate count (N_{gate}) can be used. This interconnect-gate relationship is formulated as follows,

$$X = \alpha k N_{\text{gate}} \left(1 - N_{\text{gate}}^{p-1}\right) \tag{8.9}$$

where α is calculated by Eq. 8.4.

As illustrated in Fig. 8.2, if a planar 2D design is partitioned into two separate dies whose gate counts are N_1 and N_2, respectively, by using Eq. 8.9, the interconnect count on each layer, X_1 and X_2, can be calculated, respectively, as follows,

$$X_1 = \alpha k_1 N_1 \left(1 - N_1^{p_1-1}\right) \tag{8.10}$$

$$X_2 = \alpha k_2 N_2 \left(1 - N_2^{p_2-1}\right). \tag{8.11}$$

Table 8.1: The parameters used in the metal layer estimation model

Rent's exponent for logic, p_{logic}	0.63
Rent's coefficient for logic, k_{logic}	0.14
Rent's exponent for memory, p_{memory}	0.12
Rent's coefficient for memory, k_{memory}	6.00
Average gate fanout, $f.o.$	4
Utilization efficiency of layer 1, η_1	9%
Utilization efficiency of layer 2, η_2	25%
Utilization efficiency of layer 3, η_3	54%
Utilization efficiency of layer 4, η_4	38%
Utilization efficiency of other odd layers, $\eta_{5,7,9,11,\ldots}$	60%
Utilization efficiency of other even layers, $\eta_{6,8,10,12,\ldots}$	42%
Wire pitch of lower metal layers, $w_{i(i<9)}$	$(i+1)F$
Wire pitch of higher metal layers, $w_{i(i\geq9)}$	$10F$
Blockage area per via on layer i, Av_i	$(w_i+3F)^2$

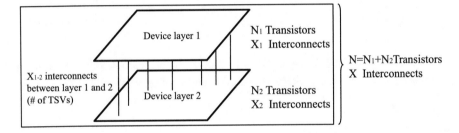

Figure 8.2: The basic idea of how to estimate the number of TSVs [111].

Irrespective of the partition pattern, the total interconnect count of a certain design always keeps constant. Thus, the number of TSVs can be calculated as follows,

$$X_{TSV} = \alpha k_{1,2}(N_1 + N_2)\left(1 - (N_1 + N_2)^{p_{1,2}-1}\right)$$
$$-\alpha k_1 N_1\left(1 - N_1^{p_1-1}\right) - \alpha k_2 N_2\left(1 - N_2^{p_2-1}\right) \tag{8.12}$$

where $k_{1,2}$ and $p_{1,2}$ are the equivalent Rent's coefficient and exponent, and they are derived as follows [114],

$$p_{1,2} = \frac{p_1 N_1 + p_2 N_2}{N_1 + N_2} \tag{8.13}$$

$$k_{1,2} = \left(k_1^{N_1} k_2^{N_2}\right)^{1/(N_1+N_2)} \tag{8.14}$$

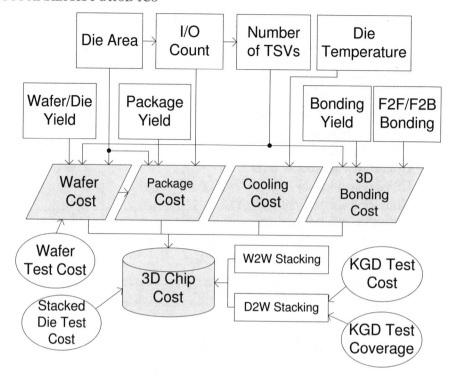

Figure 8.3: Overview of the proposed 3D cost model, which includes four components: wafer cost model, 3D bonding cost model, package cost model, and cooling cost model [111].

where N_i is the gate count on layer i.

Therefore, after adding the area overhead caused by TSVs, we formulate the new TSV-included area estimation, A_{3D}, as follows:

$$A_{3D} = A_{die} + N_{TSV/die} A_{TSV} \tag{8.15}$$

where A_{die} is calculated by die area estimator, $N_{TSV/die}$ is the equivalent number of TSVs on each die, and A_{TSV} is the size of TSVs (including the keepout area). In practice, the final TSV-included die area, A_{3D}, is used in the wafer cost estimation later, while A_{die} is used in Eq. 8.7 for metal layer estimation because that is the actually die area available for routing.

After estimating the die area of each layer, the number of required metal layers, and the number of TSVs across planar dies in the previous section, a complete set of cost models are proposed in this section. Our 3D IC cost estimation is composed of four separate parts: wafer cost model, 3D bonding cost model, package cost model, and cooling cost model. An overview of this proposed cost model set is illustrated in Fig. 8.3.

Wafer Cost Model.   The wafer cost model estimates the cost of each separate planar die. This cost estimation includes all the cost incurred before the 3D bonding process. Before estimating the die cost, we first model the wafer cost.

The wafer cost is determined by several parameters, such as fabrication process type (i.e., CMOS logic or DRAM memory), process node (from 180nm down to 22nm), wafer diameter (200mm or 300mm), and the number of metal layers (polysilicon, aluminum, or copper depending on the fabrication process type). We model the wafer cost by dividing it into two parts: a fixed part that is determined by process type, process node, wafer diameter, and the actual fabrication vendor; a variable part that is mainly contributed to the number of metal layers. We call the fixed part *silicon cost* and the variable part *metal cost*. Therefore, the wafer cost is expressed as follows:

$$C_{\text{wafer}} = C_{\text{silicon}} + N_{\text{metal}} \times C_{\text{metal}} \tag{8.16}$$

where C_{wafer} is the wafer cost, C_{silicon} is the silicon cost (the fixed part), C_{metal} is the metal cost per layer, and N_{metal} is the number of metal layers. All the parameters used in this model are from the ones collected by *IC Knowledge LLC* [115], which considers several factors including material cost, labor cost, foundry margin, number of reticles, cost per reticle, and other miscellaneous cost. Figure 8.4 shows the predicted wafer cost of 90*nm*, 65*nm*, and 45*nm* processes, with 9 or 10 layers of metal, for three different foundries, respectively.

Besides the number of metal layers affecting the cost as demonstrated by Eq. 8.16, the die area is another key factor that affects the bare die cost since smaller die area implies more dies per wafer. The number of dies per wafer is formulated by Eq. 8.17 [116] as follows,

$$N_{\text{die}} = \frac{\pi \times (\phi_{\text{wafer}}/2)^2}{A_{\text{die}}} - \frac{\pi \times \phi_{\text{wafer}}}{\sqrt{2 \times A_{\text{die}}}} \tag{8.17}$$

where N_{die} is the number of dies per wafer, ϕ_{wafer} is the diameter of the wafer, and A_{die} is the die area. After having estimated the wafer cost by Eq. 8.16 and the number of dies per wafer by Eq. 8.17, the bare die cost can be expressed as follows,

$$C_{\text{die}} = \frac{C_{\text{wafer}}}{N_{\text{die}}}. \tag{8.18}$$

In addition, the die area also relates to the die yield, which later affects the net die cost. By assuming rectangular defect density distribution, the relationship between the die area and die yield can be formulated as follows [117],

$$Y_{\text{die}} = Y_{\text{wafer}} \times \frac{\left(1 - e^{-2A_{\text{die}}D_0}\right)}{2A_{\text{die}}D_0} \tag{8.19}$$

where Y_{die} and Y_{wafer} are the yields of dies and wafers, respectively, and D_0 is the wafer defect density. Therefore, the net die cost is $C_{\text{die}}/Y_{\text{die}}$.

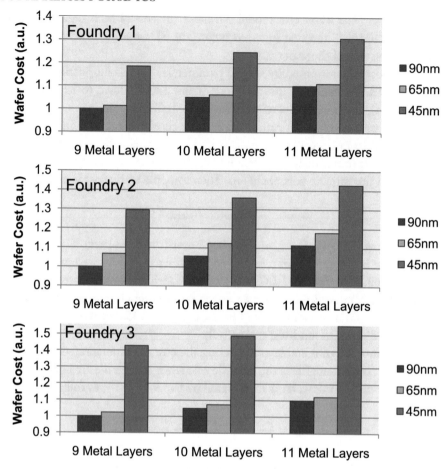

Figure 8.4: A batch of data calculated by the wafer cost model [111]. The wafer cost varies from different processes, different number of metal layers, different foundries, and some other factors.

3D Bonding Cost Model.   The 3D bonding cost model estimated the cost incurred during the process that integrates several planar 2D dies together using the TSV-based technology. The extra fabrication steps required by 3D integrations consist of TSV forming, thinning, and bonding.

There are two ways to build 3D TSVs: *laser drilling or etching*. Laser drilling is only suitable for a small number of TSVs (hundreds to thousands) while etching is suitable for a large number of TSVs. Furthermore, there are two approaches for TSV etching: (1) *TSV-first approach:* TSVs can be formed during the 2D die fabrication process, before the Back-End-of-Line (BEOL) processes. Such an approach is called *TSV-first* approach, and is shown in Fig. 8.5(a); (2) *TSV-last approach:* TSVs can also be formed after the completion of 2D fabrications, after the BEOL processes. Such an approach is called *TSV-later* approach, and is shown in Fig. 8.5(b). Either

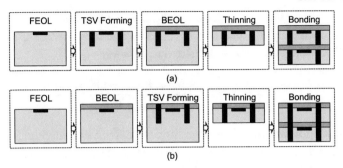

Figure 8.5: Fabrication steps for 3D ICs: (a) TSVs are formed before BEOL process, thus TSVs only punch through the silicon substrate but not the metal layers; (b) TSVs are formed after BEOL process, thus TSvs. punch through not only the silicon substrate but the metal layers as well [111].

approach has advantages and disadvantages. *TSV-first* approach builds TSVs before metal layers, thus there are no TSVs punching through metal layers and hence no TSV area overhead; *TSV-last* approach has TSV area overhead, but it isolates the die fabrication from 3D bonding, which does not need to change the traditional fabrication process. In order to separate the wafer cost model and the 3D bonding cost model, we assume *TSV-last* approach is used in 3D IC fabrication. The TSV area overhead caused by *TSV-last* approach is modeled by Eq. 8.15. The data of the 3D bonding cost including the cost of TSV forming, wafer thinning, and wafer (or die) bonding are obtained from our industry partner, with the assumption that the yield of each 3D process step is 99%.

Combined with the wafer cost model, the 3D bonding cost model can be used to estimate the cost of a 3D-stacked chip with multiple dies. In addition, the entire 3D-stacked chip cost depends on some design options. For instance, it depends on whether die-to-wafer (D2W) or wafer-to-wafer (W2W) bonding is used, and it also depends on whether face-to-face (F2F) or face-to-back (F2B) bonding is used. If D2W bonding is selected, cost of Known-Good-Die (KGD) test should also be included [118].

For D2W bonding, the cost of a bare N-layer 3D-stacked chip before packaging is calculated as follows,

$$C_{\text{D2W}} = \frac{\sum_{i=1}^{N} \left(C_{\text{die}_i} + C_{\text{KGDtest}} \right) / Y_{\text{die}_i} + (N-1) C_{\text{bonding}}}{Y_{\text{bonding}}^{N-1}} \qquad (8.20)$$

where C_{KGDtest} is the KGD test cost, which we model as $C_{\text{wafersort}} / N_{\text{die}}$, and the wafer sort cost, $C_{\text{wafersort}}$, is a constant value for a specific process in a specific foundry in our model. Note that the testing cost for 3D IC itself is a complicated problem. Our other study has demonstrated a more detailed test cost analysis model with the study of design-for-test (DFT) circuitry and various testing strategies, showing the variants of testing cost estimations [119]. For example, adding extra DFT circuitry to improve the yield of each die in D2W stacking can help the cost reduction but the increased area may increase the cost. In this book, we adopt the above testing cost assumptions to simplify the total cost estimation, due to space limitation.

For W2W bonding, the cost is calculated as follows,

$$C_{\text{W2W}} = \frac{\sum_{i=1}^{N} C_{\text{die}_i} + (N-1)C_{\text{bonding}}}{\left(\Pi_{i=1}^{N} Y_{\text{die}_i}\right) Y_{\text{bonding}}^{N-1}}. \tag{8.21}$$

In order to support multiple-layer bonding, the default bonding mode is F2B. If F2F mode is used, there is one more component die that does not need the thinning process, and the thinning cost of this die is subtracted from the total cost.

Package Cost Model.   The package cost model estimates the cost incurred during the packaging process, which can be determined by three factors: package type, package area, and pin count.

As an example, we select the package type to be flip-chip Land Grid Array (fcLGA) in this book. fcLGA is the common package type of multiprocessors, while other types of packages, such as fcPGA, PGA, pPGA, etc., are also available in our model. Besides package type, package area and pin count are the other two key factors that determine the package cost. By using the early stage estimation methodology as mentioned in Sec. 8.1, the package area can be estimated from the die area, A_{3D} in Eq. 8.15, and the pin count can be estimated by *Rent's Rule* in Eq. 8.2. We analyze the actual package cost data and find that the number of pins becomes the dominant factor to the package cost when the die area is much smaller than the total area of pin pads. It is easy to understand, since there is a base material and production cost per pin which will not be reduced with die area shrinking.

Figure 8.6 shows the sample data of package cost obtained from *IC Knowledge LLC* [115]. Based on this set of data, we use curve fitting to derive a package cost function with the parameters of the die area and the pin count, which can be expressed as follows,

$$C_{\text{package}} = \mu_1 N_p + \mu_2 A_{\text{3D}}^{\alpha} \tag{8.22}$$

where N_p is the pin count, A_{3D} is the die area. In addition, μ_1, μ_2, and α are the coefficients and exponents, which are 0.00515, 0.141 and 0.35, respectively. By observing Eq. 8.22, we can also find that pin count will dominate the package cost if the die area is sufficiently small, since the order of pin count is higher than that of area. Figure 8.6 also shows the curve fitting result, which is well aligned to the raw data.

Cooling Cost Model.   Although the combination of the aforementioned wafer cost model, 3D bonding cost model, and package cost model can already offer the cost of a 3D-stacked chip after packaging, we add the cooling cost model because it has been widely noticed that 3D-stacked chip has higher working temperature than its 2D counterpart and it might need a more powerful cooling solution that causes extra cost.

Gunther, et al. [120] noticed that the cooling cost is related to the power dissipation of the system. Furthermore, the system power dissipation is highly related to the chip working tem-

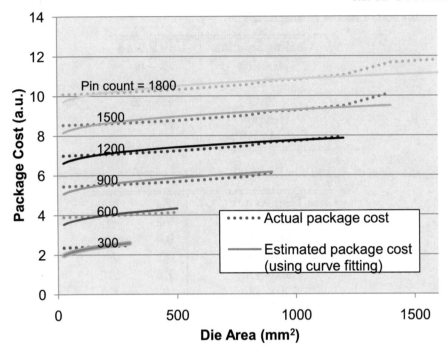

Figure 8.6: The package cost depends on both pin count and die area [111].

perature [69]. Our cooling cost estimation is based on the peak steady state temperature of the targeted 3D-stacked multiprocessor.

There are many types of cooling solutions, ranging from a simple extruded aluminum heatsink to an elaborate vapor phase refrigeration. Depending on the cooling mechanisms, cooling solutions used today can be defined as either convection cooling, phase-change cooling, thermoelectric cooling (TEC), or liquid cooling [121]. Typical convection cooling solutions are heatsinks and fans, which are widely adopted for the microprocessor chips in today's desktop computers. Phase-change cooling solutions, such as heatpipes, might be used in laptop computers. In addition, thermoelectric cooling and liquid cooling are used in some high-end computers. We collect the cost of all the cooling solutions from the commercial market by searching the data from *Digikey* [122] and *Heatsink Factory* [123]. We find that more powerful types of cooling solutions often lead to higher costs as expected. Table 8.2 is a list of typical prices of these cooling solutions.

In our model, we further assume that cooling cost increases linearly with the rise of chip temperature, if the same type of cooling solution is adopted. Based on this assumption and the data listed in Table 8.2, the cooling cost is therefore estimated as follows:

$$C_{\text{cooling}} = K_c T + c \tag{8.23}$$

Table 8.2: The cost of various cooling solutions

Cooling Solution	Unit Price $
Heatsink	3 – 6
Fan	10 – 40
TEC	15 – 20
Heatpipe	30 – 70
Liquid Cooling	75–

Table 8.3: The values of K_c and c in Eq. (8.23), which are related to the chip temperature

Chip Temperature (°C)	K_c	c
< 60	0.2	−6
60 – 90	0.4	−16
90 – 120	0.2	2
120 – 150	1.6	−170
150 – 180	2	−230

where T is the temperature from which the cooling solution can bring down, K_c and c are the cooling cost parameters, which can be determined by Table 8.3.

Figure 8.7 shows the cost of these five types of cooling solutions. It can be observed that the chips with higher steady state temperatures will require more powerful cooling solutions, which lead to higher costs. It is also illustrated in Fig. 8.7 that the cooling cost is not a global linear function of the temperature, whereas there are several regions with linear increase of the cost. Each of the regions is correspondent to one type of cooling solution.

8.2 COST EVALUATION FOR MANY-CORE MICROPROCESSOR DESIGNS

In the previous section, we evaluated the cost benefit of 3D fabrication for fully customized ASIC designs. However, the gate-level fine-granularity 3D partitioning is not feasible in the near future since it requires all the basic modules (such as adders, multipliers, etc.) to be redesigned in a 3D-partitioned way. Hence, as a near-term goal, the module-level coarse-granularity 3D partitioning is regarded as a more realistic way for 3D IC implementation, such as system-on-chip (SoC) style or many-core microprocessor design style,[4] where each IP block (such as each core/memory IP) remains to be a 2D block and the 3D partitioning is performed at the block level. In such design style, the Rent's rule-based TSV estimation is not valid, and the TSV estimation would have to be carried out based on system-level specification (such as the signal connections between cores

---

[4]In this book, a "multi-core" chip refers to eight or less homogeneous cores, whereas a "many-core" chip has more than eight homogeneous cores in one microprocessor package.

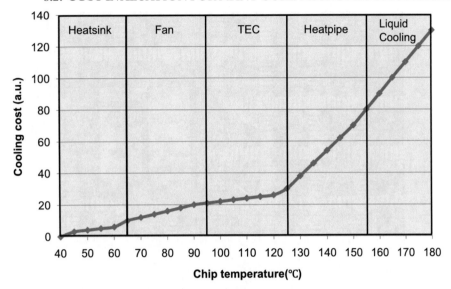

Figure 8.7: A plot for the cooling cost model: the cooling cost increases linearly with the chip temperature if the same cooling solution is used; more powerful cooling solutions result in higher costs [111].

and memory). In this section, we demonstrate how to use the remaining parts of the proposed 3D cost model to estimate the 3D many-core microprocessor cost. We use 45nm IBM common platform cost model in this case study.

Baseline 2D Configuration.   In order to have a scalable architecture for many-core microprocessors, the baseline 2D design used in our study is an NoC-based architecture, in which the processing tile is the basic building block. Each processing tile is composed of one in-order SPARC-like microprocessor core [112], one L1 cache, one L2 cache, and one router. Figure 8.8 shows the example of such architecture with 64 tiles connected by an 8-by-8 mesh structure. Table 8.4 lists the detail of cores, L1, L2 caches, and routers, respectively. The gate count of the core and the router modules is obtained from empirical data of existing designs [112], and the cell count of the cache module by only considering the SRAM cells and ignoring the cost of peripheral circuits. In this case study, we simply assume logic gates and SRAM cells (6-transistor model) occupy the same die area. We consider the resulting estimation error to be tolerable since it is an early design stage estimation. Note that we set the memory cell count (from L1 and L2 caches) to be close to the logic gate count (from cores and routers) by tuning cache capacities. This manual tuning is just for the convenience of our later heterogeneous stacking. Generally, there are no constraints on the ratio between the logic and the memory modules.

Figure 8.9 illustrates two ways to implement a 3D dual-core microprocessor, and it can be conceptually extended to the many-core microprocessor case. Generally, the two types of parti-

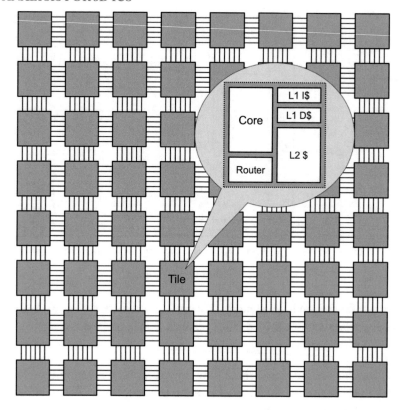

Figure 8.8: Structure of the baseline 2D multiprocessor, which consists of 64 tiles [111].

tioning strategies are: *homogeneous* and *heterogeneous*. As shown in Fig. 8.9(a), all the layers after *homogeneous* partitioning are identical, while *heterogenous* partitioning leads to core (logic) layers and cache (memory) layers as shown in Fig. 8.9(b).

8.2.1   COST EVALUATION WITH HOMOGENEOUS PARTITIONING

We first explore the *homogeneous* partitioning strategies, in which all the processing tiles are distributed into 1, 2, 4, 8, and 16 layers, respectively, without breaking tiles into finer granularity.

Figure 8.10 illustrates the conceptual view of a 4-layer homogeneous stacking. The role of TSVs in the homogeneous 3D-stacked chip is to provide the inter-layer NoC connection and to deliver the power from I/O pins to inner layers. Instead of using Eq. 8.12, the TSV count is predicted on the basis of NoC bus width. We assume 3D mesh is used and the NoC flit size is 128 bits. Therefore, the number of TSVs between two layers, N_{TSV}, is estimated as,

$$N_{TSV} = 2 \times 128 \times N_{tiles/layer} \div 50\% \tag{8.24}$$

Table 8.4: The configuration of the SPARC-like core in a tile of the baseline 2D multiprocessor

Processor Cores	
Clock frequency	1.2 GHz
Architecture type	in-order
INT pipeline	6 stages
FP pipeline	6 stages
ALU count	1
FPU count	1
Empirical gate count	1.7 million
Routers	
Type	4 ports
Empirical gate count	1.0 million
Caches	
Cache line size	64B
L1 I-cache capacity	32 KB
L1 D-cache capacity	32 KB
L2 cache capacity	256 KB
Empirical cell count	2.6 million

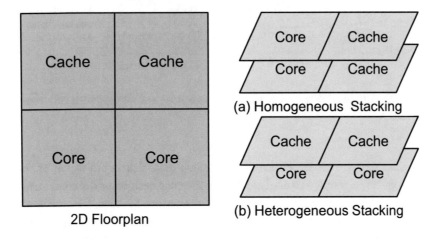

Figure 8.9: Three ways to implement a 3D dual-core microprocessor [111]

where $N_{\text{tiles/layer}}$ is the number of tiles per layer, 2 means the bi-directional channel between two tiles, and 50% is added due to our assumption that half TSVs are for power delivery. After calculation, the TSV area overhead is around 3.8%.

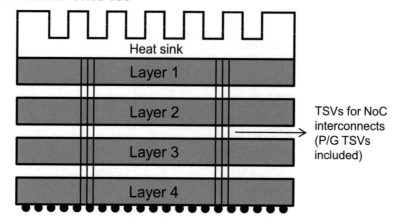

Figure 8.10: The conceptual view of a 4-layer homogeneous stacking composed of 4 identical layers [111].

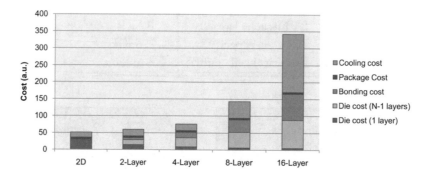

Figure 8.11: The cost breakdown of a 16-core microprocessor design with different homogeneous partitioning [111].

In the next step, we use the cost model set to estimate the cost of wafer, 3D bonding, package, and cooling, respectively. We analyze microprocessor designs of different scales. Figure 8.11 to Fig. 8.14 show the estimated price breakdown of 16-core, 32-core, 64-core, and 128-core microprocessors when they are fabricated by 1-layer, 2-layer, 4-layer, 8-layer, and 16-layer processes, respectively. In these figures, all the prices are divided into five parts (the first bar, which represents 2D planar process, only has three parts as the 2D process only has one die and does not need the bonding process). From bottom up, they are the cost of one die, the cost of remaining dies, the bonding cost, the package cost, and the cooling cost.

As the figures illustrate, the cost of one die is descending when the number of layers is ascending. This descending trend is mainly due to the yield improvement brought by the smaller die size. However, this trend is flattened at the end since the yield improvement is not significant

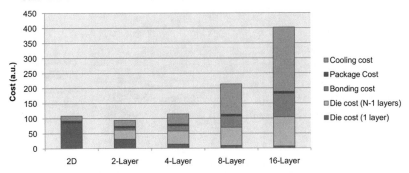

Figure 8.12: The cost breakdown of a 32-core microprocessor design with different homogeneous partitioning [111].

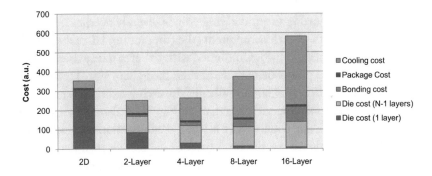

Figure 8.13: The cost breakdown of a 64-core microprocessor design with different homogeneous partitioning [111].

any more when the die size is sufficiently small. As a result, the cumulative cost of all the dies does not follow a descending trend. Combined with the 3D bonding cost, which is an ascending function to the layer count, the wafer cost usually hits its minimum at some middle points. For example, if only considering the wafer cost and the 3D bonding cost, the optimal number of layers for 16-core, 32-core, 64-core, and 128-core designs are 2, 4, 4, 8, respectively.

However, the cost before packaging is not the final cost. System-wise, the final cost needs to include the package cost and the cooling cost. As mentioned, the package cost is mainly determined by the pin count. Thus, the package cost almost remains constant in all the cases since the pin count does not change from one partitioning to another partitioning. The cooling cost mainly depends on the chip temperature. As more layers are vertically stacked, the higher the chip temperature is, the cooling cost grows with the increase of 3D layers, and this growth is quite fast because a 3D-stacked chip with more than eight layers can easily reach the peak temperature of more than 150°C. Actually, the extreme cases in this experiment (such as 16-layer

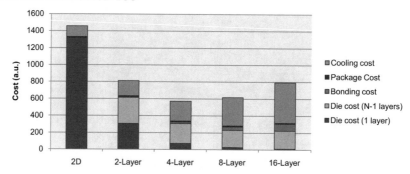

Figure 8.14: The cost breakdown of a 128-core microprocessor design with different homogeneous partitioning [111].

stacking) is not practical indeed as the underlying chip is severely overheated. In this experiment, we first obtain the power data of each component (the core, router, L1 cache, and L2 cache) from a power estimation tool, McPAT [103]. Combined with the area data achieved from early design stage estimation, we get the power density data, then feed them into HotSpot [121], which is a 3D-aware thermal estimation tool, and finally get the estimated chip temperature and its corresponding cost of cooling solutions. Figure 8.11 to Fig. 8.14 also show the cost of packaging and cooling. As we can find, the cooling cost is just a small portion for few layer stacking (such as 2-layer), but it starts to dominate the total cost for aggressive stacking (such as 8-layer and 16-layer). Therefore, the optimal partitioning solution in terms of the minimum total cost differs from the one in terms of the cost only before packaging. Observed from the result, the optimal number of layers for 16-core, 32-core, 64-core, and 128-core become 1, 2, 2, and 4, respectively instead. This result also reveals our motivation for why the package cost and the cooling cost are included in the decision of optimal 3D stacking layers.

8.2.2   COST EVALUATION WITH HETEROGENEOUS PARTITIONING

Similar to our baseline design illustrated in Fig. 8.8, today's high-performance microprocessors have a large portion of the silicon area occupied by on-chip SRAM or DRAM caches. In addition, non-volatile memory can also be integrated as on-chip memory [124]. However, different fabrication processes are not compatible. For example, the CMOS logic module might require 1-poly-9-copper-1-aluminum interconnect layers, while DRAM memory module needs 7-poly-3-copper, and flash memory module needs 4-poly-1-tungsten-2-aluminum. As a result, integrating heterogeneous modules in a planar 2D chip would dramatically increase the chip cost. Intel shows that heterogeneous integration for large 2D System-on-Chip (SoC) would boost the chip cost by three times [125].

However, fabricating heterogeneous modules separately could be a cost-effective way for such systems. For example, if the SRAM cache modules are fabricated separately, as per our es-

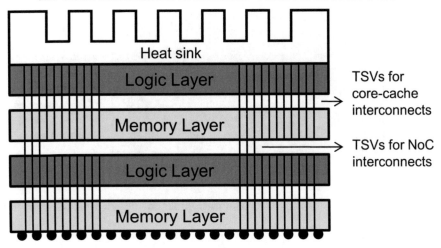

Figure 8.15: The conceptual view of a 4-layer heterogeneous stacking composed of two logic layers and two memory layers [111].

timation, the required number of metals is only 6 while the logic modules (such as cores and routers) usually need more than 10 layers of metals. Hence, We reevaluate the many-core microprocessor cost by using *heterogeneous integration*, in which the tile is broken into logic parts (i.e., core and router) and memory parts (i.e., L1 and L2 caches).

Figure 8.15 illustrates a conceptual view of 4-layer heterogeneous partitioning where logic modules (cores and routers) and memory modules (L1 and L2 caches) are on separate dies. As shown in Fig. 8.15, there are two types of TSVs in the heterogeneous stacking chip: one is for NoC mesh interconnect, which is the same as that in the homogeneous stacking case; the other is for the interconnect between cores and caches, which is caused by the separation of the logic and memory modules. The number of the extra core-to-cache TSVs is calculated as follows,

$$N_{\text{extraTSV}} = \left(\text{Data} + \sum \text{Address}_i\right) \times N_{\text{tiles/layer}} \tag{8.25}$$

where Data is the cache line size and Address_i is the address width of cache i. Note that the number of tiles per layer, $N_{\text{tiles/layer}}$ is doubled in the heterogeneous partitioning compared to its homogeneous counterpart. For our baseline configuration (Table 8.4), in which IL1 is 32KB, DL2 is 32KB, L2 is 256KB, and cache line size is 64B, $\text{Data} + \sum \text{Address}_i$ is 542. As a side-effect, the extra core-to-cache TSVs cause the TSV area overhead increasing from 3.8% to 15.7%.

Compared to its homogeneous counterpart, the advantages and disadvantages of partitioning an NoC-based many-core microprocessor in the heterogeneous way are,

- Fabricating logic and memory dies separately reduces the wafer cost;

- Putting logic dies closer to heat sink reduces the cooling cost;

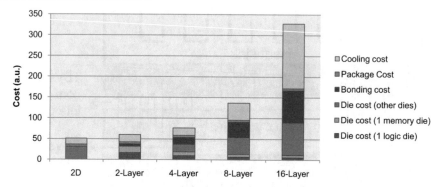

Figure 8.16: The cost breakdown of a 16-core microprocessor design with different heterogeneous partitioning [111].

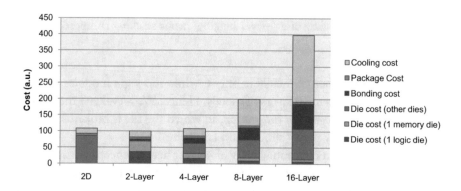

Figure 8.17: The cost breakdown of a 32-core microprocessor design with different heterogeneous partitioning [111].

- The higher TSV area overhead increases the wafer cost and the package cost.

In order to know whether heterogenous integration is more cost-effective, we repeat the cost estimation for 16-core, 32-core, 64-core, and 128-core microprocessor designs by using heterogenous partitioning. The price breakdown of heterogeneous stacking is shown in Fig. 8.16 to Fig. 8.19. Compared to the homogeneous integration cost as illustrated in Fig. 8.11 to Fig. 8.14, we can observe that the cost of heterogeneous integration is higher than that of homogenous integration in most cases, which is mainly due to the increased die area caused by higher TSV counts. However, in some cases, the cost of heterogenous integration can be cheaper compared to its homogeneous counterparts. Figure 8.20 demonstrates the cost comparison between homogeneous and heterogeneous integrations for a 64-core microprocessor design using the 45nm 4-layer 3D process. It can be observed that although the fabrication efficiency of separated logic and memory

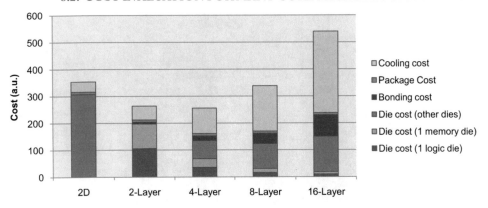

Figure 8.18: The cost breakdown of a 64-core microprocessor design with different heterogeneous partitioning [111].

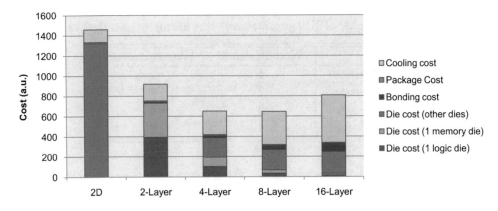

Figure 8.19: The cost breakdown of a 128-core microprocessor design with different heterogeneous partitioning [111].

dies cannot offset the extra cost caused by the increased die area, the cooling cost is reduced after putting logic layers, which closer to the heat sink (or other cooling solutions).

While our experiment shows heterogeneous integration has the cost advantages on cheaper memory layer and lower chip temperature, in reality, some designers might still prefer homogeneous integration, which has identical layout on each die. The reciprocal design symmetry (RDS) was proposed by Alam, et al. [126] to re-use one mask for multiple 3D-stacked layers. The RDS could relieve the design effort, and thus reduce the design time, which means a cost reduction on human resources. However, this personnel cost is not included in our cost model set, and it is out of the scope of this book.

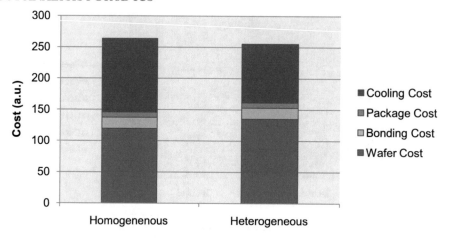

Figure 8.20: The cost comparison between homogeneous and heterogeneous integrations for a 64-core microprocessor using 45nm 2-layer 3D process [111].

Table 8.5: The optimal number of 3D layers for many-core microprocessor designs (45nm)

	Homogeneous	Heterogeneous
16-core	1	1
32-core	2	2
64-core	2	4
128-core	4	8

We list the optimal partitioning options for 45nm many-core microprocessor designs in Table 8.5, which shows how our proposed cost estimation methodology helps decide the 3D partitioning at the early design stage.

Conclusion. To overcome the barriers in technology scaling, a three-dimensional integrated circuit (3D IC) is emerging as an attractive option for future microprocessor designs. However, fabrication cost is one of the important considerations for the wide adoption of the 3D integration. System-level cost analysis at the early design stage to help the decision making on whether 3D integration should be used for the application is very critical.

To facilitate the early design stage cost analysis, we propose a set of cost models that include wafer cost, 3D bonding cost, package cost, and cooling cost. Based on the cost analysis, we identify the design opportunities for cost reduction in 3D ICs, and provide a few design guidelines on cost-effective 3D IC designs. Our research is complementary to the existing research on 3D microprocessors that are focused on other design goals, such as performance and power consumption.

CHAPTER 9

Conclusion

In this book, we have reviewed the background of emerging 3D integration technologies and explored various architecture designs that employ 3D integration. 3D integration technologies promise high performance, low power, low cost, and high density microprocessor architecture solutions. It is an attractive solution in developing high-performance, energy-efficient, thermal-aware, and cost-effective chip-multiprocessors and GPU systems. In particular, with chip-multiprocessors, 3D integration provides low wire latency in connecting processor cores and caches. With GPU systems, 3D integration is promising in developing high-bandwidth, low power graphics memory interface. 3D integration also enlarges the capacity of on-chip memory, which can be employed as the last-level cache, a portion of main memory, or the combination of both.

Although 3D integration brings in great benefits and opportunities in architecture design, several challenges exist on the way toward its wide adoption in future computer systems. This book has discussed two of them—thermal and cost—and explored thermal-aware and cost-aware microprocessor design techniques. Beyond these two challenges, the following two design challenges are also critical towards the wide adoption of 3D integration technologies.

Design tools and methodologies.   3D integration technologies will not be commercially viable without the support of electronic design automation (EDA) tools and methodologies. Given particular design goals, efficient EDA tools and methodologies can help architects and circuit designers make decisions on whether to adopt 3D or 2D integration. They can also help designers address design trade-offs in performance, power, and cost, when 3D integration is adopted. Furthermore, 3D ICs may require new layout rules that might be driven by features on adjacent die; with larger sizes compared with conventional vias, TSvs. can illustrate significant new layout feature; power planning in 3D ICs still requires substantial studies due to the complexity introduced by the third dimension; efficient EDA tools need to take thermal constraints into consideration; place and route tools need to consider thermal constraints of 3D ICs to avoid hot spots; efficient analysis tools are required for handling electromagnetic interference concern for 3D ICs. To efficiently exploit the benefits of 3D technologies, design tools and methodologies that support 3D integration are imperative [6].

Testing methodologies.   In 3D IC design, various testing strategies and integration methods can affect system performance, power, and cost dramatically [127]. An obstacle towards the adoption of 3D technologies is therefore insufficient understanding of 3D testing issues and the lack of design-for-testability (DFT) techniques for 3D ICs. Without considering test during the design phase, effective tests on 3D ICs will become impossible. It is therefore very crucial to study options in 3D testing methodologies. Both new standards and design tools are required to diagnose issues properly. 3D IC fabrication incorporates many more intermediate steps than conventional 2D IC fabrication, such as die stacking and TSV bonding. These extra steps require wafer tests before final assembly and packaging. However, a wafer test for 3D ICs is challenging in three ways. First, existing probe technology cannot perform finer pitch and dimensions of TSV tips and is limited to handling only several hundred probes at a much lower number than required TSV probes. Second, a wafer test may require creating a known-good die (KGD) stack, which may be at the risk of damaging due to the contact of the highly thinned wafer by a wafer probe. 3D ICs can also impose intra-die defects, which may be caused by thermal issues and new manufacturing steps such as wafer thinning and bonding the top of a TSV to another wafer. Consequently, new fault models are required to address this issue. While 3D IC testing is a crucial problem, it has remained largely unexplored in the research community and requires substantial efforts in exploring them.

Bibliography

[1] J. Zhao, C. Xu, and Y. Xie, "Bandwidth-aware reconfigurable cache design with hybrid memory technologies," in *Proceedings of the International Conference on Computer-Aided Design (ICCAD)*, 2011, pp. 48–55. DOI: 10.1109/ICCAD.2011.6105304. xi

[2] "MICRON, micron collaborates with Intel to enhance Knights landing with a high performance, on-package memory solution," http://investors.micron.com/releasedetail.cfm?ReleaseID=856057. xii

[3] "NVIDIA," http://blogs.nvidia.com/blog/2013/03/19/nvidia-ceo-updates-nvidias-roadmap/. xii, 45

[4] "AMD, what is heterogeneous system architecture (HSA)?" http://developer.amd.com/resources/heterogeneous-computing/what-is-heterogeneous-system-architecture-hsa/. xii

[5] Y. Xie, G. Loh, B. Black, and K. Bernstein, "Design space exploration for 3D architectures," *ACM Journal of Emerging Technologies in Compuing Systems*, 2006. DOI: 10.1145/1148015.1148016. 2, 7, 10, 16, 53, 65, 73

[6] Y. Xie, J. Cong, and S. Sapatnekar, *Three-Dimensional Integrated Circuit Design: EDA, Design and Microarchitectures*. Springer, 2009. DOI: 10.1007/978-1-4419-0784-4. 2, 97

[7] P. Garrou, *Handbook of 3D Integration: Technology and Applications using 3D Integrated Circuits*. Wiley-CVH, 2008, ch. Introduction to 3D Integration. 2, 3

[8] W. R. Davis, J. Wilson, S. Mick, J. Xu, H. Hua, C. Mineo, A. M. Sule, M. Steer, and P. D. Franzon, "Demystifying 3D ICs: the pros and cons of going vertical," *IEEE Design and Test of Computers*, vol. 22, no. 6, pp. 498– 510, 2005. DOI: 10.1109/MDT.2005.136. 3

[9] J. Burns, G. Carpenter, E. Kursun, R. Puri, J. Warnock, and M. Scheuermann, "Design, cad and technology challenges for future processors: 3d perspectives," in *Design Automation Conference (DAC), 2011 48th ACM/EDAC/IEEE*, June 2011, pp. 212–212. DOI: 10.1145/2024724.2024772. 4, 5

[10] J. Joyner, P. Zarkesh-Ha, and J. Meindl, "A stochastic global net-length distribution for a three-dimensional system-on-a-chip (3D-SoC)," in *Proc. 14th Annual IEEE International ASIC/SOC Conference*, Sep. 2001. DOI: 10.1109/ASIC.2001.954688. 7

[11] Y.-f. Tsai, F. Wang, Y. Xie, N. Vijaykrishnan, and M. J. Irwin, "Design space exploration for three-dimensional cache," *IEEE Transactions on Very Large Scale Integration Systems*, 2008. DOI: 10.1109/ASIC.2001.954688. 7, 10

[12] B. Vaidyanathan, W.-L. Hung, F. Wang, Y. Xie, V. Narayanan, and M. J. Irwin, "Architecting microprocessor components in 3D design space," in *Intl. Conf. on VLSI Design*, 2007, pp. 103–108. DOI: 10.1109/VLSID.2007.41. 7, 8, 13, 23

[13] K. Puttaswamy and G. H. Loh, "Scalability of 3D-integrated arithmetic units in high-performance microprocessors," in *Design Automation Conference*, 2007, pp. 622–625. DOI: 10.1109/DAC.2007.375238.

[14] J. Ouyang, G. Sun, Y. Chen, L. Duan, T. Zhang, Y. Xie, and M. Irwin, "Arithmetic unit design using 180nm TSV-based 3D stacking technology," in *IEEE International 3D System Integration Conference*, 2009. DOI: 10.1109/3DIC.2009.5306565. 8

[15] R. Egawa, J. Tada, H. Kobayashi, and G. Goto, "Evaluation of fine grain 3D integrated arithmetic units," in *IEEE International 3D System Integration Conference*, 2009. DOI: 10.1109/3DIC.2009.5306566. 7, 8

[16] B. Black et al., "Die stacking 3D microarchitecture," in *MICRO*, 2006, pp. 469–479. DOI: 10.1109/MICRO.2006.18. 8, 9

[17] G. H. Loh, "3d-stacked memory architectures for multi-core processors," in *International Symposium on Computer Architecture (ISCA)*, 2008, pp. 453–464. DOI: 10.1145/1394608.1382159. 8, 9

[18] P. Jacob et al., "Mitigating memory wall effects in high clock rate and multi-core cmos 3D ICs: Processor memory stacks," *Proceedings of IEEE*, vol. 96, no. 10, 2008.

[19] G. Loh, Y. Xie, and B. Black, "Processor design in three-dimensional die-stacking technologies," *IEEE Micro*, vol. 27, no. 3, pp. 31–48, 2007. DOI: 10.1109/MM.2007.59. 8

[20] T. Kgil, S. D'Souza, A. Saidi, N. Binkert, R. Dreslinski, T. Mudge, S. Reinhardt, and K. Flautner, "PicoServer: using 3D stacking technology to enable a compact energy efficient chip multiprocessor," in *ASPLOS*, 2006, pp. 117–128. DOI: 10.1145/1168857.1168873. 9, 37

[21] S. Vangal et al., "An 80-tile Sub-100-W TeraFLOPS processor in 65-nm CMOS," *IEEE Journal of Solid-State Circuits*, vol. 43, no. 1, pp. 29–41, 2008. DOI: 10.1109/JSSC.2007.910957. 9

[22] G. Loh, "Extending the effectiveness of 3D-stacked dram caches with an adaptive multi-queue policy," in *International Symposium on Microarchitecture (MICRO)*, Dec. 2009. DOI: 10.1145/1669112.1669139. 9

[23] F. Li, C. Nicopoulos, T. Richardson, Y. Xie, N. Vijaykrishnan, and M. Kandemir, "Design and management of 3D chip multiprocessors using network-in-memory," in *International Symposium on Computer Architecture (ISCA'06)*, 2006. DOI: 10.1145/1150019.1136497. 10

[24] X. Dong, X. Wu, G. Sun, Y. Xie, H. Li, and Y. Chen, "Circuit and microarchitecture evaluation of 3D stacking Magnetic RAM (MRAM) as a universal memory replacement," in *Design Automation Conference*, 2008, pp. 554–559. DOI: 10.1145/1391469.1391610. 10

[25] X. Wu, J. Li, L. Zhang, E. Speight, and Y. Xie, "Hybrid cache architecture," in *International Symposium on Computer Architecture (ISCA)*, 2009. DOI: 10.1145/1555815.1555761. 10, 13

[26] G. Sun, X. Dong, Y. Xie, J. Li, and Y. Chen, "A novel 3D stacked MRAM cache architecture for CMPs," in *International Symposium on High Performance Computer Architecture*, 2009. DOI: 10.1109/HPCA.2009.4798259. 10

[27] D. Vantrease, R. Schreiber, M. Monchiero, M. McLaren, N. P. Jouppi, M. Fiorentino, A. Davis, N. Binkert, R. G. Beausoleil, and J. H. Ahn, "Corona: System implications of emerging nanophotonic technology," in *Proceedings of the 35th International Symposium on Computer Architecture*, 2008, pp. 153–164. DOI: 10.1109/ISCA.2008.35. 11

[28] X. Dong and Y. Xie, "Cost analysis and system-level design exploration for 3D ICs," in *Asia and South Pacific Design Automation Conference*, 2009. DOI: 10.1145/1509633.1509700. 11

[29] F. Li, C. Nicopoulos, T. Richardson, Y. Xie, V. Narayanan, and M. Kandemir, "Design and management of 3D chip multiprocessors using network-in-memory," in *ISCA '06: Proceedings of the 33rd annual international symposium on Computer Architecture*. Washington, DC, USA: IEEE Computer Society, 2006, pp. 130–141. DOI: 10.1145/1150019.1136497. 13, 50

[30] G. H. Loh, "Extending the effectiveness of 3D-stacked DRAM caches with an adaptive multi-queue policy," in *MICRO 42: Proceedings of the 42nd Annual IEEE/ACM International Symposium on Microarchitecture*. New York, NY, USA: ACM, 2009, pp. 201–212. DOI: 10.1145/1669112.1669139. 13, 27, 28, 29

[31] G. H. Loh, "3D-stacked memory architectures for multi-core processors," in *ISCA '08: Proceedings of the 35th Annual International Symposium on Computer Architecture*. Washington, DC, USA: IEEE Computer Society, 2008, pp. 453–464. DOI: 10.1145/1394608.1382159. 29, 30

[32] D. H. Woo, N. H. Seong, D. L. Lewis, and H.-H. S. Lee, "An optimized 3D-stacked memory architecture by exploiting excessive, high-density TSV bandwidth," in *HPCA '10: Proceedings of the 2007 IEEE 16th International Symposium on High Performance Computer Architecture*. Bangalore, Indian: IEEE Computer Society, 2010. DOI: 10.1109/HPCA.2010.5416628. 31

[33] M. Ghosh and H.-H. S. Lee, "Smart Refresh: An enhanced memory controller design for reducing energy in conventional and 3D die-stacked DRAMs," in *MICRO 40: Proceedings of the 40th Annual IEEE/ACM International Symposium on Microarchitecture*. Washington, DC, USA: IEEE Computer Society, 2007, pp. 134–145. DOI: 10.1109/MICRO.2007.38. 13, 29, 32

[34] K. Puttaswamy and G. H. Loh, "3D-integrated SRAM components for high-performance microprocessors," *IEEE Trans. Comput.*, vol. 58, no. 10, pp. 1369–1381, 2009. DOI: 10.1109/TC.2009.92. 13, 20, 21, 22

[35] B. Black, M. Annavaram, N. Brekelbaum, J. DeVale, L. Jiang, G. H. Loh, D. McCaule, P. Morrow, D. W. Nelson, D. Pantuso, P. Reed, J. Rupley, S. Shankar, J. Shen, and C. Webb, "Die stacking (3D) microarchitecture," in *MICRO 39: Proceedings of the 39th Annual IEEE/ACM International Symposium on Microarchitecture*. Washington, DC, USA: IEEE Computer Society, 2006, pp. 469–479. DOI: 10.1109/MICRO.2006.18. 22, 23, 24, 27, 28

[36] K. Puttaswamy and G. H. Loh, "Thermal herding: Microarchitecture techniques for controlling hotspots in high-performance 3D-integrated processors," in *HPCA '07: Proceedings of the 2007 IEEE 13th International Symposium on High Performance Computer Architecture*. Washington, DC, USA: IEEE Computer Society, 2007, pp. 193–204. DOI: 10.1109/ICCD.2004.1347939. 25, 26

[37] B. Black, D. W. Nelson, C. Webb, and N. Samra, "3D processing technology and its impact on iA32 microprocessors," in *ICCD '04: Proceedings of the IEEE International Conference on Computer Design*. Washington, DC, USA: IEEE Computer Society, 2004, pp. 316–318. 13

[38] G. Sun, X. Dong, Y. Xie *et al.*, "A novel architecture of the 3D stacked MRAM L2 cache for CMPs," in *Proceedings of the International Symposium on High Performance Computer Architecture*, 2009, pp. 239–249. DOI: 10.1109/HPCA.2009.4798259. 13

[39] K. Zhang, U. Bhattacharya, Z. Chen, F. Hamzaoglu, D. Murray, N. Vallepalli, B. Z. Y. Wang, and M. Bohr, "A sram design on 65nm cmos technology with integrated leakage reduction scheme," in *VLSI Technology Digest of Technical Papers*, 2004. DOI: 10.1109/VLSIC.2004.1346592. 14

[40] Y. Tsai, Y. Xie, V. Narayanan, and M. J. Irwin, "Three-dimensional cache design explo-ration using 3DCacti," *Proceedings of the IEEE International Conference on Computer Design (ICCD 2005)*, pp. 519–524, 2005. DOI: 10.1109/ICCD.2005.108. 15, 17, 18, 19, 20, 53

[41] P. Shivakumar *et al.*, "Cacti 3.0: An Integrated Cache Timing, Power, and Area Model," *Western Research Lab Research Report*, 2001. 15

[42] Y. Ma, Y. Liu, E. Kursun, G. Reinman, and J. Cong, "Investigating the effects of fine-grain three-dimensional integration on microarchitecture design," *J. Emerg. Technol. Comput. Syst.*, vol. 4, pp. 17:1–17:30, November 2008. [Online]. Available: http://doi.acm.org/10.1145/1412587.1412590 DOI: 10.1145/1324177.1324180. 24

[43] T. Kgil, A. Saidi, N. Binkert, S. Reinhardt, K. Flautner, and T. Mudge, "Picoserver: Using 3D stacking technology to build energy efficient servers," *J. Emerg. Technol. Comput. Syst.*, vol. 4, no. 4, pp. 1–34, 2008. DOI: 10.1145/1412587.1412589. 29, 31

[44] C. Liu, I. Ganusov, M. Burtscher, and S. Tiwari, "Bridging the processor-memory per-formance gap with 3D IC technology," *Design Test of Computers, IEEE*, vol. 22, no. 6, pp. 556 – 564, nov. 2005. DOI: 10.1109/MDT.2005.134.

[45] G. L. Loi, B. Agrawal, N. Srivastava, S.-C. Lin, T. Sherwood, and K. Banerjee, "A thermally-aware performance analysis of vertically integrated (3-D) processor-memory hi-erarchy," in *DAC '06: Proceedings of the 43rd annual Design Automation Conference.* New York, NY, USA: ACM, 2006, pp. 991–996. DOI: 10.1145/1146909.1147160. 29

[46] T. Semiconductors, "Leo FaStack 1Gb DDR SDRAM datasheet," in *http://www.tezzaron.com/memory/TSC Leo.htm*, 2005. 29

[47] K. Chen, S. Li, N. Muralimanohar, J. H. Ahn, J. Brockman, and N. Jouppi, "CACTI-3DD: Architecture-level modeling for 3D die-stacked DRAM main memory," in *Pro-ceedings of the Design, Automation Test in Europe Conference Exhibition*, 2012, pp. 33–38. DOI: 10.1109/DATE.2012.6176428. 32

[48] P. Shivakumar, , and N. Jouppi, "Cacti 3.0: An Integrated Cache Timing, Power, and Area Model," *Western Research Lab Research Report*, 2001/2. 32

[49] X. Dong, Y. Xie, N. Muralimanohar, and N. P. Jouppi, "Simple but effective heterogeneous main memory with on-chip memory controller support," in *Proceedings of the International Conference for High Performance Computing*, 2010, pp. 1–11. DOI: 10.1109/SC.2010.50. 33

[50] "Itrs 2009." [Online]. Available: http://www.itrs.net 33

[51] "NAS parallel benchmarks," http://www.nas.nasa.gov/publications/npb.html. 33

[52] B. M. Beckmann, M. R. Marty, and D. A. Wood, "ASR: Adaptive selective replication for CMP caches," in *Proceedings of the 39th Annual IEEE/ACM International Symposium on Microarchitecture*, 2006, pp. 443–454. DOI: 10.1109/MICRO.2006.10. 34

[53] Z. Chishti, M. D. Powell, and T. N. Vijaykumar, "Optimizing replication, communication, and capacity allocation in CMPs," in *Proceedings of the 32Nd Annual International Symposium on Computer Architecture*, 2005, pp. 357–368. DOI: 10.1145/1080695.1070001.

[54] J. Huh, C. Kim, H. Shafi, L. Zhang, D. Burger, and S. W. Keckler, "A NUCA substrate for flexible CMP cache sharing," in *Proceedings of the 19th Annual International Conference on Supercomputing*, 2005, pp. 31–40. DOI: 10.1145/1088149.1088154.

[55] N. Rafique, W.-T. Lim, and M. Thottethodi, "Architectural support for operating system-driven CMP cache management," in *Proceedings of the 15th International Conference on Parallel Architectures and Compilation Techniques*, 2006, pp. 2–12. DOI: 10.1145/1152154.1152160.

[56] S. Cho and L. Jin, "Managing distributed, shared L2 caches through OS-level page allocation," in *Proceedings of the 39th Annual IEEE/ACM International Symposium on Microarchitecture*, 2006, pp. 455–468. DOI: 10.1109/MICRO.2006.31. 34

[57] G. Loh and M. Hill, "Supporting very large DRAM caches with compound-access scheduling and MissMap," pp. 70–78, May 2012. DOI: 10.1109/MM.2012.25. 34

[58] M. K. Qureshi and G. H. Loh, "Fundamental latency trade-off in architecting DRAM caches: Outperforming impractical SRAM-tags with a simple and practical design," in *Proceedings of the 45th Annual IEEE/ACM International Symposium on Microarchitecture*, ser. MICRO-45, 2012, pp. 235–246. DOI: 10.1109/MICRO.2012.30. 35

[59] C. Chou, A. Jaleel, and M. K. Qureshi, "CAMEO: A two-level memory organization with capacity of main memory and flexibility of hardware-managed cache," in *Proceedings of the 47th Annual IEEE/ACM International Symposium on Microarchitecture*, ser. MICRO-47, 2014. DOI: 10.1109/MICRO.2014.63. 36, 37

[60] S. Keckler, W. Dally, B. Khailany, M. Garland, and D. Glasco, "GPUs and the future of parallel computing," *Micro, IEEE*, vol. 31, no. 5, pp. 7–17, Sept 2011. DOI: 10.1109/MM.2011.89. 39

[61] Q. Zou, M. Poremba, R. He, W. Yang, J. Zhao, and Y. Xie, "Heterogeneous architecture design with emerging 3d and non-volatile memory technologies," in *Design Automation Conference (ASP-DAC), 2015 20th Asia and South Pacific*, Jan 2015, pp. 785–790. DOI: 10.1109/ASPDAC.2015.7059106. 39

[62] Y.-J. Lee and S. K. Lim, "On gpu bus power reduction with 3d ic technologies," in *Proceedings of the Conference on Design, Automation & Test in Europe*, ser. DATE '14, 2014, pp. 175:1–175:6. DOI: 10.7873/DATE.2014.188. 39

[63] AMD, "Amd radeon™hd 7970 graphics," 2012. [Online]. Available: `http://www.amd.com/us/products/desktop/graphics/7000/7970/Pages/radeon-7970.aspx` 39

[64] NVIDIA, "Quadro 6000 - workstation graphics card for 3d design, styling, visualization, cad, and more," 2010. [Online]. Available: `http://www.nvidia.com/object/product-quadro-6000-us.html` 39, 41

[65] J. Zhao, G. Sun, G. Loh, and Y. Xie, "Optimizing GPU energy efficiency with 3D die-stacking graphics memory and reconfigurable memory interface," *ACM Transactions on Architecture and Code Optimization (TACO)*, vol. 10, no. 4, pp. 24:1–24:25, 2013. DOI: 10.1145/2541228.2555301. 39

[66] E. Vick, S. Goodwin, G. Cunnigham, and D. S. Temple, "Vias-last process technology for thick 2.5d si interposers," in *3D Systems Integration Conference*, 2012, pp. 1–4. DOI: 10.1109/3DIC.2012.6262990. 40

[67] G. H. Loh, "3d-stacked memory architectures for multi-core processors," in *Proceedings of the International Symposium on Computer Architecture*, 2008, pp. 453–464. DOI: 10.1145/1394608.1382159. 40

[68] Hynix, "Hynix gddr5 sgram datasheet," 2009. [Online]. Available: `http://www.hynix.com/products/graphics/` 41

[69] K. Skadron, M. R. Stan, K. Sankaranarayanan, W. Huang, S. Velusamy, and T. D., "Temperature-aware microarchitecture: Modeling and implementation," *ACM Trans. on Architecture and Code Optimization*, vol. 1, no. 1, pp. 94–125, 2004. DOI: 10.1145/980152.980157. 41, 85

[70] A. Al Maashri, G. Sun, X. Dong, V. Narayanan, and Y. Xie, "3d gpu architecture using cache stacking: Performance, cost, power and thermal analysis," in *Proceedings of the International Conferenece on Computer Design*, 2009, pp. 254–259. DOI: 10.1109/ICCD.2009.5413147. 45

[71] J. Kim, C. Nicopoulos, D. Park, R. Das, Y. Xie, N. Vijaykrishnan, and C. Das, "A novel dimensionally-decomposed router for on-chip communication in 3D architectures," in *Proceedings of the Annual International Symposium on Computer Architecture*, 2007. DOI: 10.1145/1273440.1250680. 48

[72] P. Dongkook, S. Eachempati, R. Das, A. K. Mishra, Y. Xie, N. Vijaykrishnan, and C. R. Das, "MIRA: A multi-layered on-chip interconnect router architecture," in

International Symposium on Computer Architecture (ISCA), 2008, pp. 251–261. DOI: 10.1109/ISCA.2008.13. 48, 52, 54

[73] Y. Xu et al., "A low-radix and low-diameter 3D interconnection network design," in *Intl. Symp. on High Performance Computer Architecture*, 2009. DOI: 10.1109/HPCA.2009.4798234. 48, 55

[74] L. Carloni, P. Pande, and Y. Xie, "Networks-on-chip in emerging intercoonect paradigms: Advantages and challenges," in *Intl. Symp. on Networks-on-chips*, 2009. DOI: 10.1109/NOCS.2009.5071456. 48

[75] A. Jantsch and H. Tenhunen, *Networks on Chip*.   Kluwer Academic Publishers, 2003. 48

[76] G. De Micheli and L. Benini, *Networks on Chips*.   San Francisco, CA: Morgan Kaupmann, 2006. 48

[77] F. Li, C. Nicopoulos, T. Richardson, Y. Xie, V. Narayanan, and M. Kandemir, "Design and management of 3D chip multiprocessors using network-in-memory," in *Proceedings of International Symposium on Computer Architecture*, 2006, pp. 130–141. DOI: 10.1145/1150019.1136497. 48, 50

[78] J. Kim, C. Nicopoulos, D. Park, R. Das, Y. Xie, N. Vijaykrishnan, and C. Das, "A novel dimensionally-decomposed router for on-chip communication in 3D architectures," in *Proceedings of International Symposium on Computer Architecture*, 2007. DOI: 10.1145/1273440.1250680. 51, 52

[79] S. Yan and B. Lin, "Design of application-specific 3D networks-on-chip architectures," in *Proceedings of International Conference of Computer Design*, 2008, pp. 142–149. DOI: 10.1109/ICCD.2008.4751853. 56

[80] I. Loi, F. Angiolini, and L. Benini, "Developing mesochronous synchronizers to enable 3D NoCs," in *Proceedings of Design, Automation and Test in Europe Conference*, 2008, pp. 1414–1419. DOI: 10.1109/DATE.2008.4484872.

[81] I. Loi, S. Mitra, T. H. Lee, S. Fujita, and L. Benini, "A low-overhead fault tolerance scheme for tsv-based 3D network on chip links," in *Proceedings of International Conference on Computer-Aided Design*, 2008, pp. 598–602. DOI: 10.1145/1509456.1509589.

[82] V. F. Pavlidis and E. G. Friedman, "3-D topologies for networks-on-chip," *IEEE Transactions on Very Large Scale Integration (VLSI) Systems*, vol. 15, no. 10, pp. 1081–1090, 2007. DOI: 10.1109/TVLSI.2007.893649. 48

[83] J. Kim, C. Nicopoulos, D. Park, V. Narayanan, M. S. Yousif, and C. Das, "A gracefully degrading and energy-efficient modular router architecture for on-chip networks,"

in *Proceedings of International Symposium on Computer Architecture*, 2006, pp. 4–15. DOI: 10.1145/1150019.1136487. 52

[84] J. Kim, C. Nicopoulos, D. Park, R. Das, Y. Xie, V. Narayanan, M. S. Yousif, and C. R. Das, "A novel dimensionally-decomposed router for on-chip communication in 3D architectures," *SIGARCH Comput. Archit. News*, vol. 35, no. 2, pp. 138–149, 2007. DOI: 10.1145/1273440.1250680. 53

[85] D. Park, S. Eachempati, R. Das, A. K. Mishra, Y. Xie, N. Vijaykrishnan, and C. R. Das, "Mira: A multi-layered on-chip interconnect router architecture," in *ISCA '08: Proceedings of the 35th Annual International Symposium on Computer Architecture*. Washington, DC, USA: IEEE Computer Society, 2008, pp. 251–261. DOI: 10.1145/1394608.1382143. 54, 55

[86] Y. Ye, L. Duan, J. Xu, J. Ouyang, M. K. Hung, and Y. Xie, "3D optical networks-on-chip (NoC) for multiprocessor systems-on-chip (MPSoC)," sep. 2009, pp. 1 –6. DOI: 10.1109/3DIC.2009.5306588. 56

[87] X. Zhang and A. Louri, "A multilayer nanophotonic interconnection network for on-chip many-core communications," in *DAC '10: Proceedings of the 47th Design Automation Conference*. New York, NY, USA: ACM, 2010, pp. 156–161. DOI: 10.1145/1837274.1837314. 56

[88] D. Vantrease, R. Schreiber, M. Monchiero, M. McLaren, N. P. Jouppi, M. Fiorentino, A. Davis, N. Binkert, R. G. Beausoleil, and J. H. Ahn, "Corona: System implications of emerging nanophotonic technology," in *ISCA '08: Proceedings of the 35th Annual International Symposium on Computer Architecture*. Washington, DC, USA: IEEE Computer Society, 2008, pp. 153–164. DOI: 10.1145/1394608.1382135. 56

[89] S. Sapatnekar, "Addressing Thermal and Power Delivery Bottlenecks in 3D Circuits," in *Proceedings of the Asia and South Pacific Design Automation Conference, 2009.*, 2009, pp. 423–428. DOI: 10.1145/1509633.1509738. 60

[90] P. Li, L. T. Pileggi, M. Asheghi, and R. Chandra, "IC thermal simulation and modeling via efficient multigrid-based approaches," *IEEE Transactions on Computer-Aided Design of Integrated Circuits and Systems*, vol. 25, pp. 1763–1776, 2006. DOI: 10.1109/TCAD.2005.858276. 60

[91] B. Goplen and S. Sapatnekar, "Efficient Thermal Placement of Standard Cells in 3D ICs using a Force Directed Approach," *Proceedings of ICCAD*, 2003. DOI: 10.1145/996070.1009873. 61, 64

[92] W. Huang, S. Ghosh, and et al., "Hotspot: a compact thermal modeling methodology for early-stage VLSI design," *TVLSI*, vol. 14, no. 5, pp. 501–513, 2006. DOI: 10.1109/TVLSI.2006.876103. 61

[93] G. M. Link and N. Vijaykrishnan, "Thermal trends in emerging technologies," 2006, pp. 625–632. DOI: 10.1109/ISQED.2006.136. 61, 66

[94] J. Cong and Y. Zhang, "Thermal-driven multilevel routing for 3D ICs," in *Proc. Asia and South Pacific Design Automation Conf the ASP-DAC 2005*, vol. 1, 2005, pp. 121–126. DOI: 10.1109/ASPDAC.2005.1466143. 62

[95] Y. Yang, Z. Gu, C. Zhu, R. P. Dick, and L. Shang, "ISAC: Integrated space-and-time-adaptive chip-package thermal analysis," vol. 26, no. 1, pp. 86–99, 2007. DOI: 10.1109/TCAD.2006.882589. 63

[96] W. Hung, G. Link, Y. Xie, V. Narayanan, and M.J.Irwin, "Interconnect and thermal-aware floorplanning for 3D microprocessors," in *Proceedings of the International Symposium of Quality Electronic Devices*, 2006. DOI: 10.1109/ISQED.2006.77. 63, 64, 69

[97] J. Cong, J. Wei, and Y. Zhang, "A thermal-driven floorplanning algorithm for 3D ics," in *Proc. of International Conference Computer-Aided Design (ICCAD)*, 2004. DOI: 10.1145/1112239.1112292. 64, 67

[98] C. Chu and D. Wong, "A matrix synthesis approach to thermal placement," in *Proceedings of the ISPD*, 1997. DOI: 10.1145/267665.267708. 64

[99] C. Tsai and S. Kang, "Cell-level placement for improving substrate thermal distributio," in *IEEE Trans. on Computer-Aided Design of Integrated Circuits and System*, 2000. DOI: 10.1109/43.828554. 64, 68

[100] P. Shiu and S. K. Lim, "Multi-layer floorplanning for reliable system-on-package," in *Proc. of IEEE International Symposium on Circuits and Systems (ISCAS)*, 2004. DOI: 10.1109/IS-CAS.2004.1329460. 64, 66

[101] Y. Chang, Y. Chang, G.-M. Wu, and S.-W. Wu, "B*-trees: a new representation for non-slicing floorplans ," *Proceedings of the Annual ACM/IEEE Design Automation Conference*, 2000. DOI: 10.1109/DAC.2000.855354. 64, 65

[102] K. Puttaswamy and G. Loh, "Thermal herding: Microarchitecture techniques for controlling hotspots in high-performance 3D-integrated processors," in *High Performance Computer Architecture, 2007. HPCA 2007. IEEE 13th International Symposium on*, Feb. 2007, pp. 193–204. DOI: 10.1109/HPCA.2007.346197. 70

[103] S. Li, J. H. Ahn, R. D. Strong, J. B. Brockman, D. M. Tullsen, and N. P. Jouppi, "Mc-pat: An integrated power, area, and timing modeling framework for multicore and many-core architectures," in *Proceedings of the International Symposium on Microarchitecture*, 2009. DOI: 10.1145/1669112.1669172. 74, 92

[104] R. Weerasekera, L.-R. Zheng, D. Pamunuwa, and H. Tenhunen, "Extending Systems-on-Chip to the third dimension: performance, cost and technological tradeoffs," in *Proceedings of the International Conference on Computer-Aided Design*, 2007, pp. 212–219. DOI: 10.1145/1326073.1326117. 74, 76

[105] B. S. Landman and R. L. Russo, "On a pin versus block relationship for partitions of logic graphs," *IEEE Transancations on Computers*, vol. 20, no. 12, pp. 1469–1479, 1971. DOI: 10.1109/T-C.1971.223159. 75

[106] W. E. Donath, "Placement and average interconnection lengths of computer logic," *IEEE Transactions on Circuits and Systems*, vol. 26, no. 4, pp. 272–277, 1979. DOI: 10.1109/TCS.1979.1084635. 76, 78

[107] J. A. Davis, V. K. De, and J. D. Meindl, "A stochastic wire-length distribution for gigas-cale integration (GSI);ªPart I: derivation and validation," *IEEE Transactions on Electron Devices*, vol. 45, no. 3, pp. 580–589, 1998. DOI: 10.1109/16.661219. 76, 77

[108] A. B. Kahng, S. Mantik, and D. Stroobandt, "Toward accurate models of achievable routing," *IEEE Transactions on Computer-Aided Design of Integrated Circuits and Systems*, vol. 20, no. 5, pp. 648–659, 2001. DOI: 10.1109/43.920697. 76, 77

[109] P. Chong and R. K. Brayton, "Estimating and optimizing routing utilization in DSM design," in *Proceedings of the Workshop on System-Level Interconnect Prediction*, 1999, pp. 97–102. 77

[110] H. B. Bakoglu, *Circuits, Interconnections, and Packaging for VLSI*. Addison-Wesley, 1990. 77

[111] X. Dong, J. Zhao, and Y. Xie, "Fabrication cost analysis and cost-aware design space exploration for 3-D ICs," *Computer-Aided Design of Integrated Circuits and Systems, IEEE Transactions on*, vol. 29, no. 12, pp. 1959 –1972, dec. 2010. DOI: 10.1109/TCAD.2010.2062811. 78, 79, 80, 82, 83, 85, 87, 88, 89, 90, 91, 92, 93, 94, 95, 96

[112] M. Tremblay and S. Chaudhry, "A third-generation 65nm 16-core 32-thread llus 32-scout-thread CMT SPARC(R) processor," in *Proceedings of the International Solid State Circuit Conference*, 2008, pp. 82–83. DOI: 10.1109/ISSCC.2008.4523067. 78, 87

[113] G. Loh, Y. Xie, and B. Black, "Processor design in three-dimensional die-stacking technologies," *IEEE Micro*, vol. 27, no. 3, pp. 31–48, 2007. DOI: 10.1109/MM.2007.59. 77

[114] P. Zarkesh-Ha, J. Davis, W. Loh, and J. Meindl, "On a pin versus gate relationship for heterogeneous systems: heterogeneous Rent's rule," in *Proceedings of the IEEE Custom Integrated Circuits Conference*, 1998, pp. 93–96. DOI: 10.1109/CICC.1998.694914. 79

[115] IC Knowledge LLC., "IC Cost Model, 2009 Revision 0906," 2009. 81, 84

[116] J. Rabaey, A. Chandrakasan, and B. Nikolic, "Digital Integrated Circuits," *Prentice-Hall*, 2003. 81

[117] B. Murphy, "Cost-size optima of monolithic integrated circuits," *Proceedings of the IEEE*, vol. 52, no. 12, pp. 1537–1545, dec. 1964. DOI: 10.1109/PROC.1964.3442. 81

[118] L. Smith, G. Smith, S. Hosali, and S. Arkalgud, "3D: it all comes down to cost," in *Proceedings of RTI Conference of 3D Architecture for Semiconductors and Packaging*, 2007. 83

[119] Y. Chen, D. Niu, X. Dong, Y. Xie, and K. Chakrabarty, "Testing Cost Analsysis in three-dimensional (3D) integration technology," in *Technical Report, CSE Department, Penn State, http://www.cse.psu.edu/~yuanxie/3d.html*, 2010. 83

[120] S. Gunther, F. Binns, D. M. Carmean, and J. C. Hall, "Managing the impact of increasing microprocessor power consumption," *Intel Technology Journal*, vol. 5, no. 1, pp. 1–9, 2001. 84

[121] W. Huang, K. Skadron, S. Gurumurthi *et al.*, "Differentiating the roles of IR measurement and simulation for power and temperature-aware design," in *Proceedings of the International Symposium on Performance Analysis of Systems and Software*, 2009, pp. 1–10. DOI: 10.1109/ISPASS.2009.4919633. 85, 92

[122] Digikey, 2009, www.digikey.com. 85

[123] Heatsink Factory, 2009, www.heatsinkfactory.com. 85

[124] X. Dong, X. Wu, G. Sun *et al.*, "Circuit and microarchitecture evaluation of 3D stacking Magnetic RAM (MRAM) as a universal memory replacement," in *Proceedings of the Design Automation Conference*, 2008, pp. 554–559. DOI: 10.1145/1391469.1391610. 92

[125] S. Borkar, "3D technology: a system perspective," in *Proceedings of the International 3D-System Integration Conference*, 2008, pp. 1–14. 92

[126] S. M. Alam, R. E. Jones, S. Pozder, and A. Jain, "Die/wafer stacking with reciprocal design symmetry (RDS) for mask reuse in three-dimensional (3D) integration technology," in *Proceedings of the International Symposium on Quality of Electronic Design*, 2009, pp. 569–575. DOI: 10.1109/ISQED.2009.4810357. 95

[127] "Cadence, 3D ICs with TSVs – design challenges and requirements," http://www.cadence.com/rl/resources/white_papers/3dic_wp.pdf. 97

Authors' Biographies

YUAN XIE

Yuan Xie received his B.S. degree in electronic engineering from Tsinghua University, Beijing, in 1997, and his M.S. and Ph.D. degrees in electrical engineering from Princeton University in 1999 and 2002, respectively. He is currently a Professor in the Electrical and Computer Engineering department at the University of California at Santa Barbara. Before joining UCSB in Fall 2014, he was with the Pennsylvania State University from 2003 to 2014, and with IBM Microelectronic Division's Worldwide Design Center from 2002 to 2003. Prof. Xie is a recipient of the National Science Foundation Early Faculty (CAREER) award, the SRC Inventor Recognition Award, IBM Faculty Award, and several Best Paper Award and Best Paper Award Nominations at IEEE/ACM conferences. He has published more than 100 research papers in journals and refereed conference proceedings, in the area of EDA, computer architecture, VLSI circuit designs, and embedded systems. His current research projects include: three-dimensional integrated circuits (3D ICs) design, EDA, and architecture; emerging memory technologies; low power and thermal-aware design; reliable circuits and architectures; and embedded system synthesis. He is currently Associate Editor for ACM Journal of Emerging Technologies in Computing Systems (JETC), IEEE Transactions on Very Large Scale Integration Systems (TVLSI), IEEE Transactions on Computer Aided Design of Integrated Circuits (TCAD), IEEE Design and Test of Computers, IET Computers and Digital Techniques (IET CDT). He is a Fellow of IEEE.

JISHEN ZHAO

Jishen Zhao received her B.S. and M.S. degrees from Zhejiang University, and Ph.D. degree from Pennsylvania State University. She is currently an Assistant Professor at the University of California, Santa Cruz. Her research interests include a broad range of computer architecture topics with an emphasis on memory systems, high-performance computing, and energy efficiency. She is also interested in electronic design automation and VLSI design for three-dimensional integrated circuits and nonvolatile memories.